A GUIDE TO
ANATOMY & PHYSIOLOGY LAB

2nd Edition

THOMAS G. RUST, M.Ed., M.A.

TABLE OF CONTENTS

Subject	Page
Mitosis	1
Epithelium	2-4
Connective Tissue	4-14
Skeletal System	15-24
Muscle Tissue	25-27
Nervous System	28-38
Cardiovascular System	39-41
Lymphatic Organs	41-45
Integument	46-49
Digestive System	50-72
Respiratory System	73-75
Urinary System	76-78
Reproductive System	79-91
Embryology	92
Ear	93-94
Eye	95-96
Endocrine System	97-100
Cat Dissection	101-113
Fetal Pig Dissection	114-119

Additional copies may be ordered from: Southwest Educational Enterprises
10711 Auldine
San Antonio, Texas 78230
(512) 342-2297

ATTENTION: After November 1992, our new
Area Code will be:
(210) 342-2297

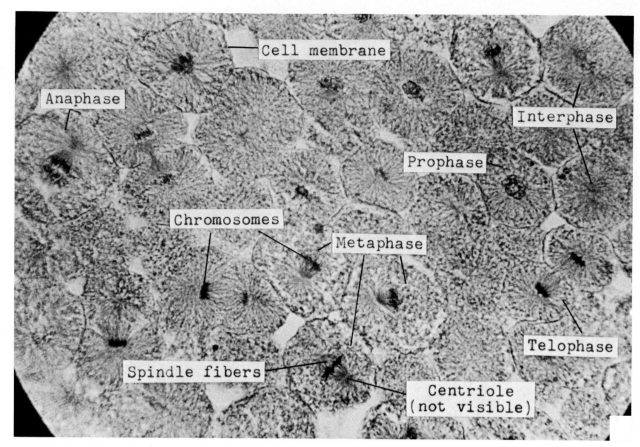

Fig. 1a   Mitosis in cells of the Whitefish blastula x.s. x430. A blastula is a hollow ball of cells formed by successive mitotic divisions of a zygote (fertilized egg). (See Fig. 92g.)

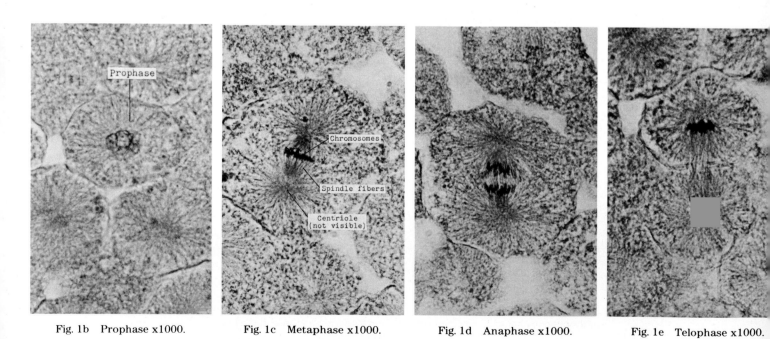

Fig. 1b   Prophase x1000.   Fig. 1c   Metaphase x1000.   Fig. 1d   Anaphase x1000.   Fig. 1e   Telophase x1000.

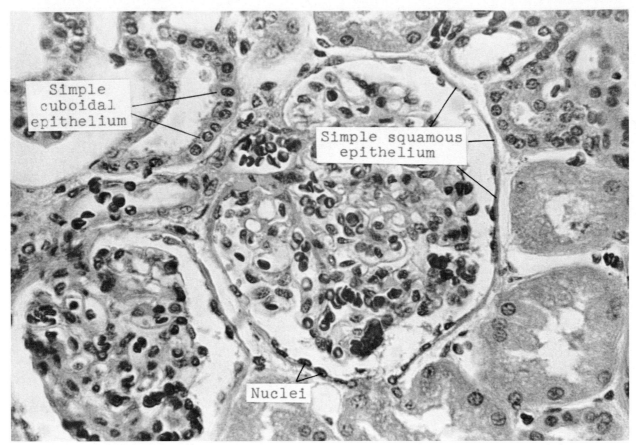

Fig. 2a Simple squamous epithelium lining Bowman's capsule. Simple cuboidal epithelium in the macula densa (kidney) x.s. x430. Also see Figs. 4b & 40d.

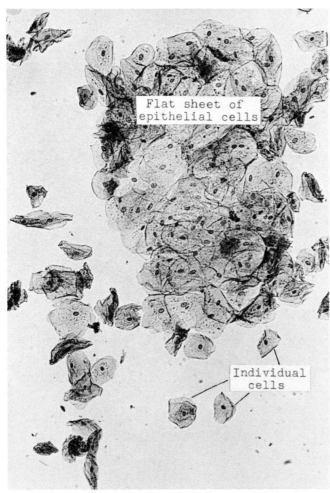

Fig. 2b Stratified squamous epithelial cells (human), scraping from the lining of the oral cavity w.m. x100.

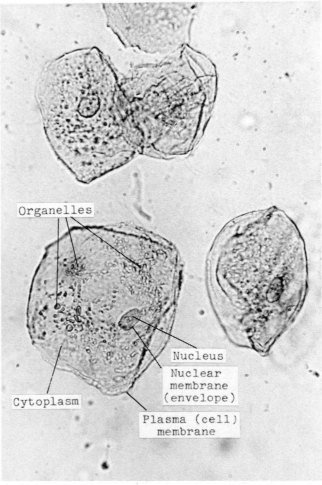

Fig. 2c Stratified squamous epithelial cells (human), scraping from the lining of the oral cavity w.m. x430.

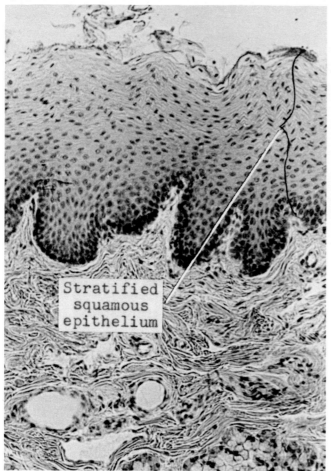

Fig. 3a   Stratified squamous epithelium lining the esophagus x.s. x100.

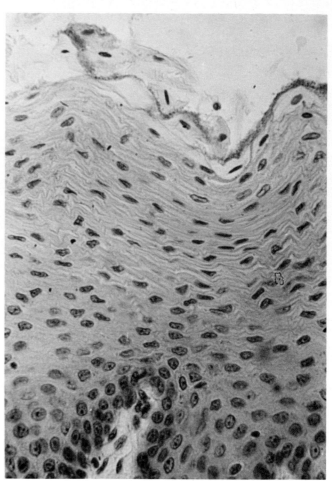

Fig. 3b   Stratified squamous epithelium lining the esophagus x.s. x430.

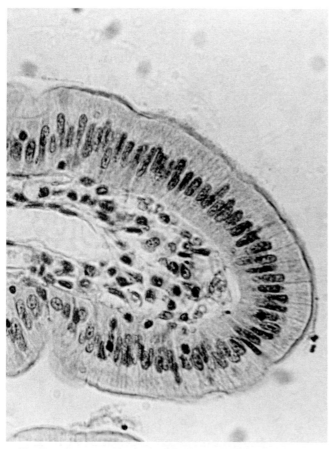

Fig. 3c   Simple columnar epithelium lining a villus of the small intestine x.s. x430.

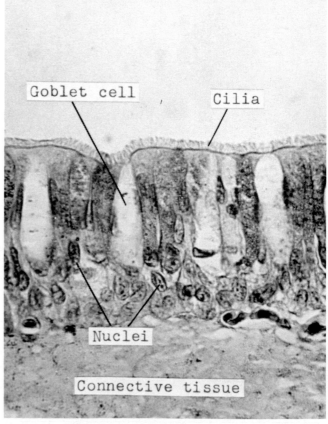

Fig. 3d   Ciliated pseudostratified columnar epithelium lining the trachea x.s. x430. (Also see Fig. 74a.)

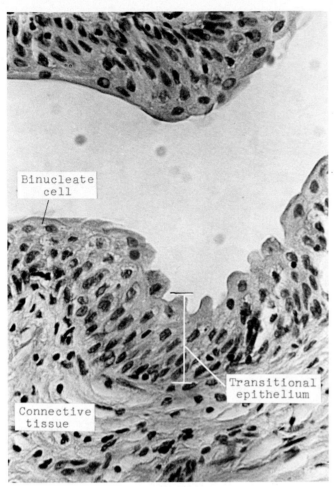

Fig. 4a   Transitional epithelium from the lining of an empty urinary bladder x.s. x430. Surface cells are usually dome shaped and often binucleate. (Also see Figs. 78c & 78d.)

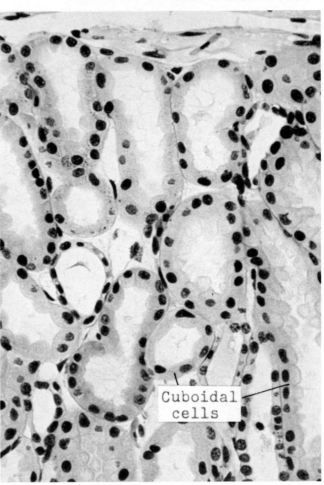

Fig. 4b   Simple cuboidal epithelium lining the distal convoluted tubules of the kidney x.s. x430. (Also see Fig. 2a.)

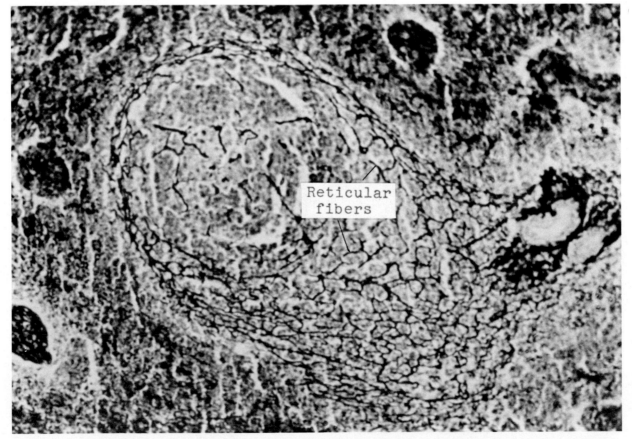

Fig. 4c   Reticular fibers in and around the germinal center of a cortical nodule in a lymph node (providing a supporting network) x.s. x100 (silver stain). (Also see Figs. 5a, 5b, 43a & 43b.)

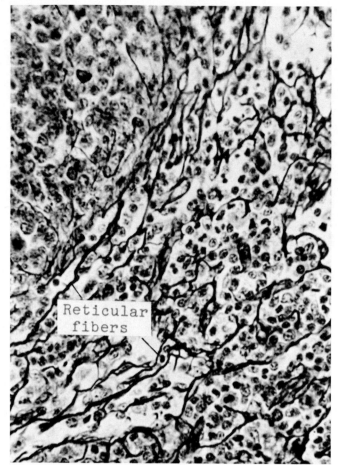

Fig. 5a   Reticular fibers in and around a cortical nodule in a lymph node x.s. x430. (Also see Figs. 4c, 43a & 43b.

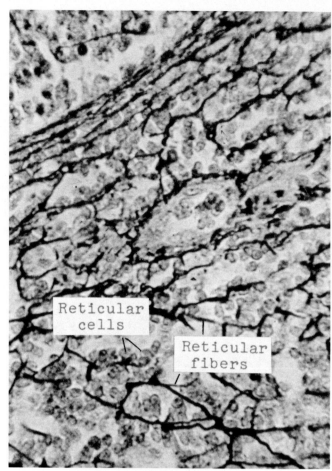

Fig. 5b   Reticular fibers and reticular cells around a cortical nodule in a lymph node x.s. x430. (Also see Figs. 4c, 43a & 43b.)

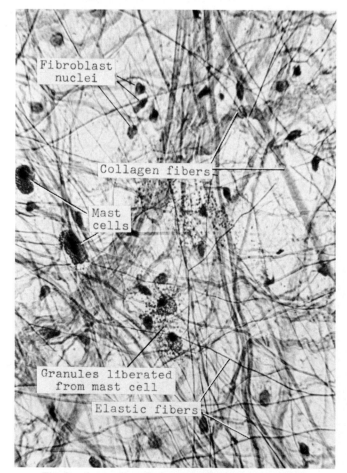

Fig. 5c   Loose (areolar) connective tissue x.s. x430.

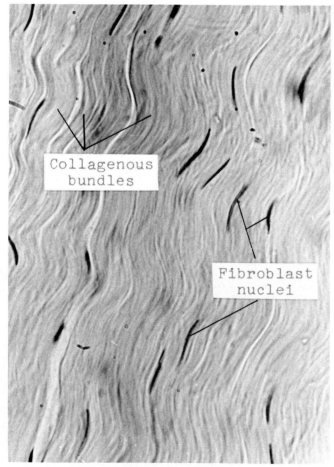

Fig. 5d   Fibrous (dense) connective tissue from the tendon l.s. x430.

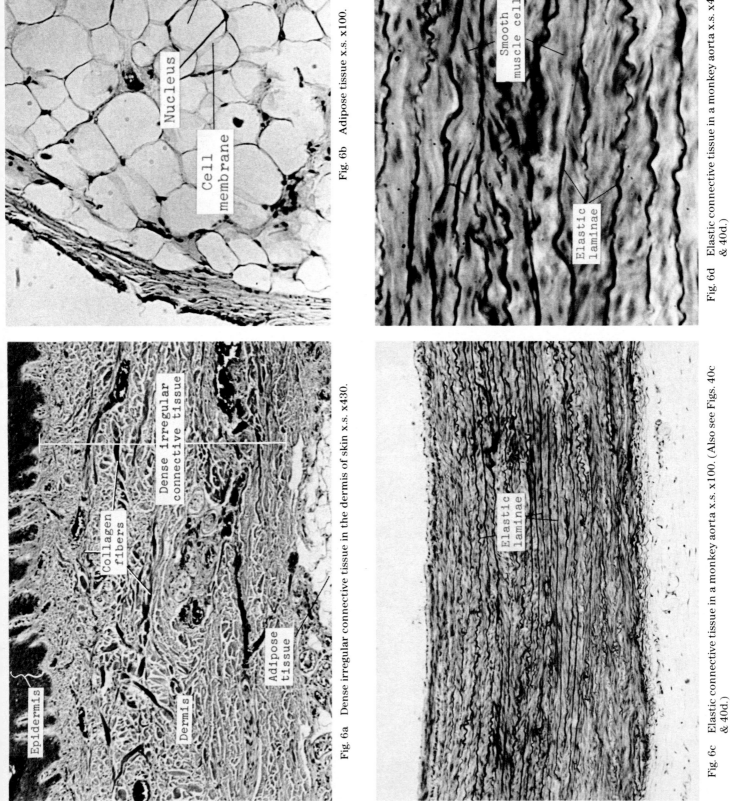

Fig. 6a Dense irregular connective tissue in the dermis of skin x.s. x430.

Fig. 6b Adipose tissue x.s. x100.

Fig. 6c Elastic connective tissue in a monkey aorta x.s. x100. (Also see Figs. 40c & 40d.)

Fig. 6d Elastic connective tissue in a monkey aorta x.s. x430. (Also see Figs. 40c & 40d.)

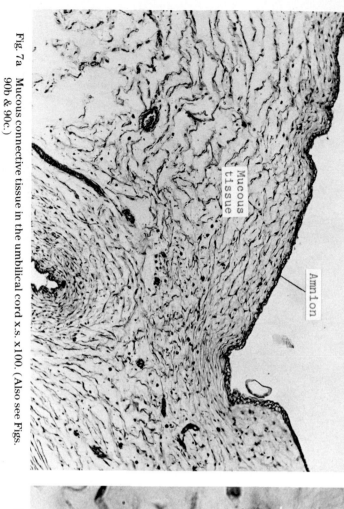

Fig. 7a   Mucous connective tissue in the umbilical cord x.s. x100. (Also see Figs. 90b & 90c.)

Mucous tissue

Amnion

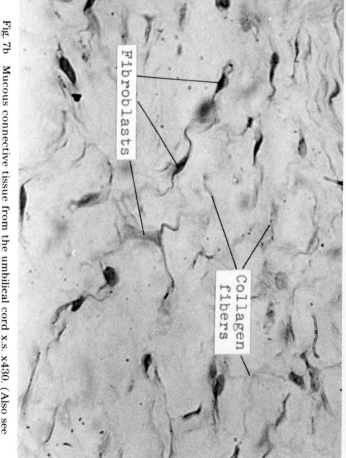

Fig. 7b   Mucous connective tissue from the umbilical cord x.s. x430. (Also see Figs. 90b & 90c.)

Fibroblasts

Collagen fibers

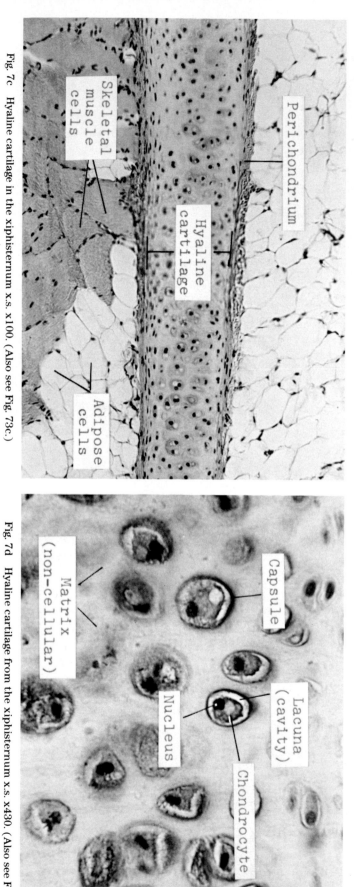

Fig. 7c   Hyaline cartilage in the xiphisternum x.s. x100. (Also see Fig. 73c.)

Skeletal muscle cells

Perichondrium

Hyaline cartilage

Adipose cells

Fig. 7d   Hyaline cartilage from the xiphisternum x.s. x430. (Also see Fig. 73c.)

Matrix (non-cellular)

Capsule

Nucleus

Lacuna (cavity)

Chondrocyte

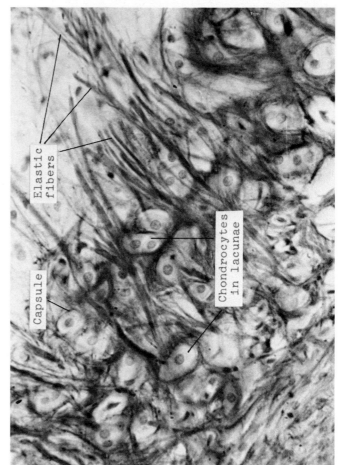

Fig. 8b Fibrocartilage from an intervertebral disc x.s. x200.

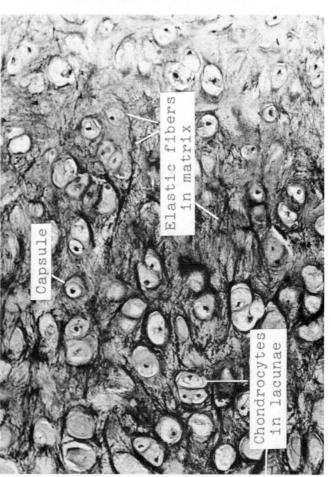

Fig. 8d Elastic cartilage x.s. x430. (Also see Fig. 94c.)

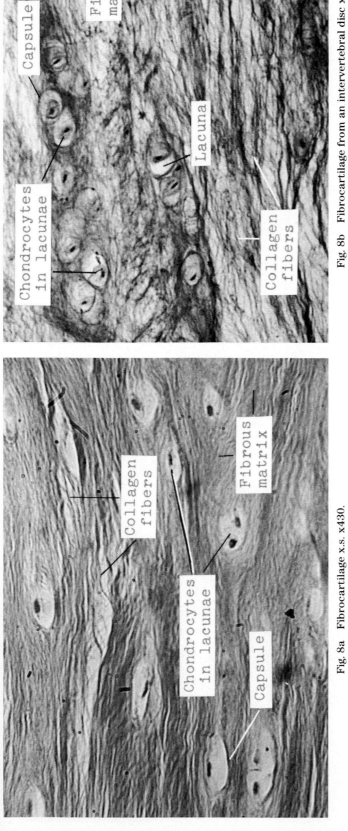

Fig. 8a Fibrocartilage x.s. x430.

Fig. 8c Elastic cartilage x.s. x100. (Also see Fig. 94c.)

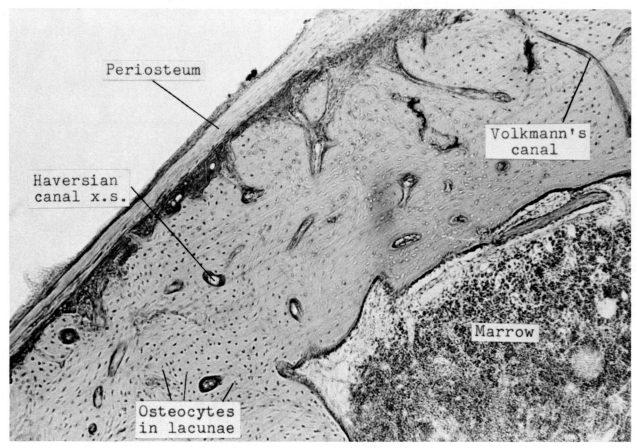

Fig. 9a   Bone x.s. x100.

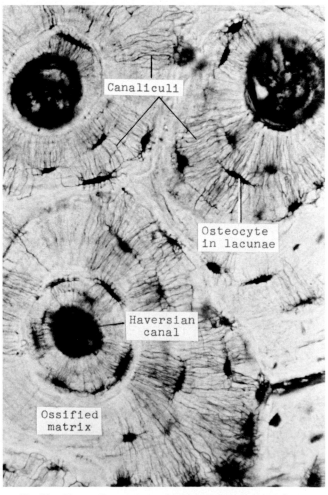

Fig. 9b   Haversian systems (osteons) in compact bone
x.s. x430.

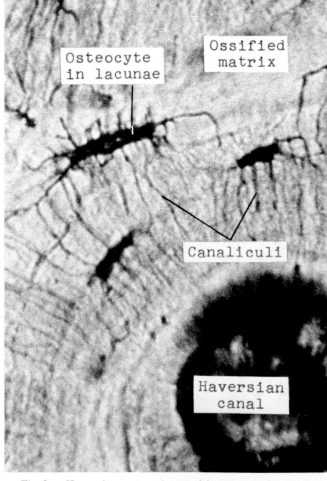

Fig. 9c   Haversian system (osteon) in compact bone x.s.
x1000.

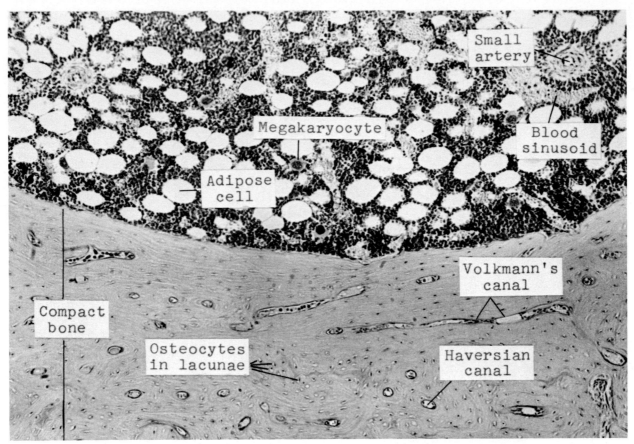

Fig. 10a   Bone marrow and compact bone x.s. x100.

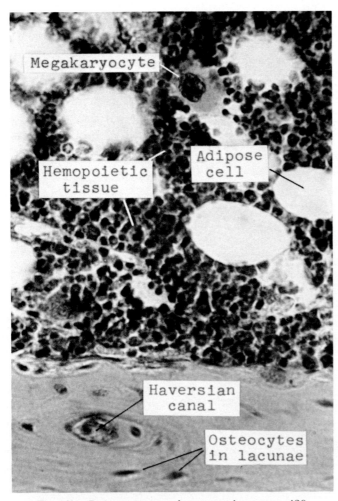

Fig. 10b   Bone marrow and compact bone x.s. x430.

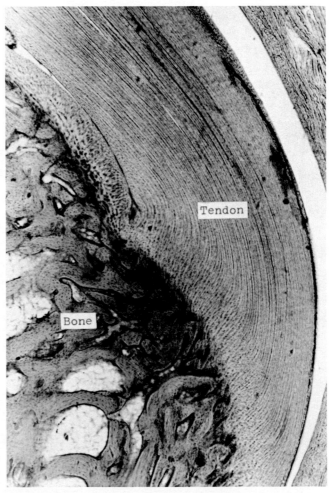

Fig. 10c   Attachment site of a tendon to bone x.s. x40.

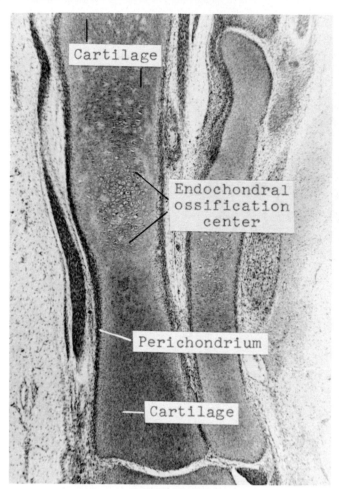

Fig. 11a   Endochondral bone development (early stage) l.s. x40.

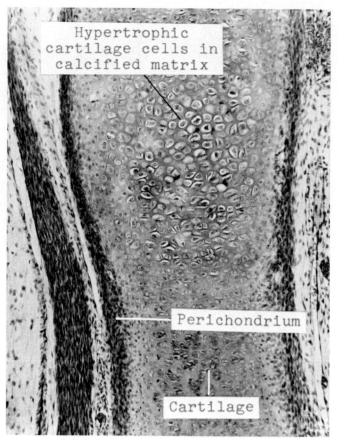

Fig. 11b   Endochondral bone development (early stage) l.s. x100.

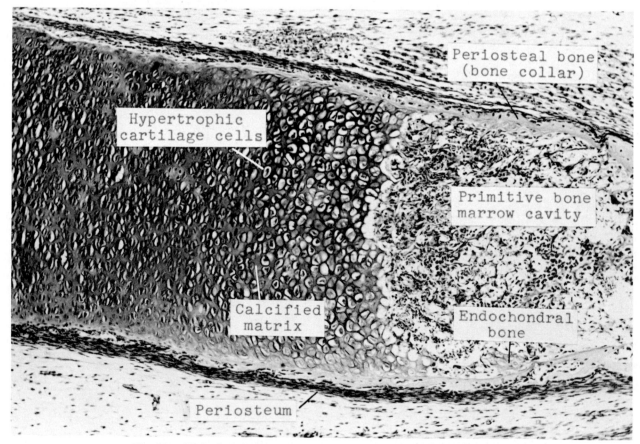

Fig. 11c   Endochondral bone development (later stage) l.s. x100.

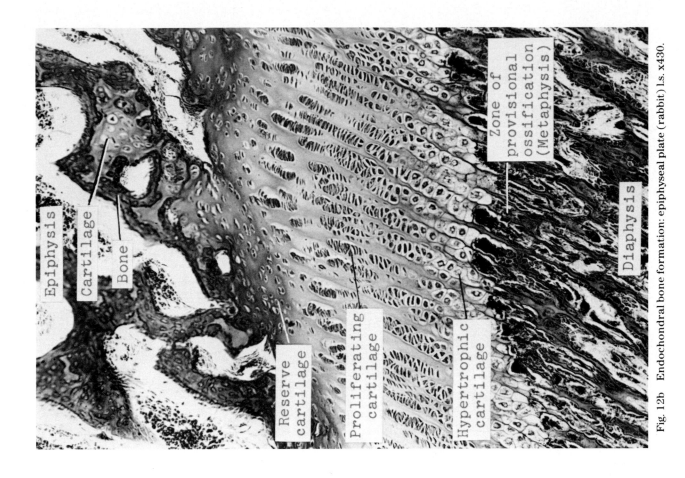

Epiphysis

Cartilage

Bone

Zone of provisional ossification (Metaphysis)

Diaphysis

Reserve cartilage

Proliferating cartilage

Hypertrophic cartilage

Fig. 12b   Endochondral bone formation: epiphyseal plate (rabbit) l.s. x430.

Secondary ossification center

Epiphyseal plate

Articular cartilage

Joint cavity

Fig. 12a   Joint and epiphyseal plate (rabbit) l.s. x20.

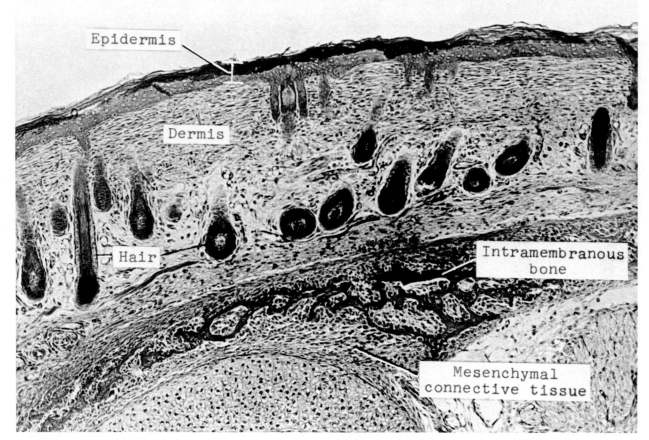

Fig. 13a   Intramembranous bone formation in the skull (early) x.s. x40.

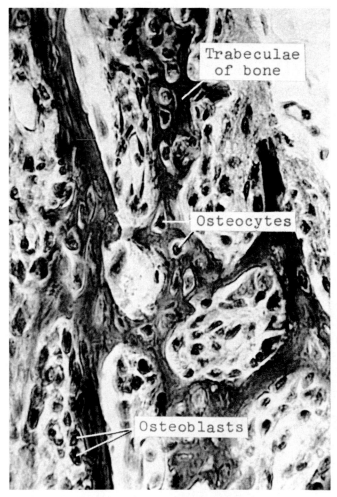

Fig. 13b   Intramembranous bone formation in the skull (early) x.s. x430.

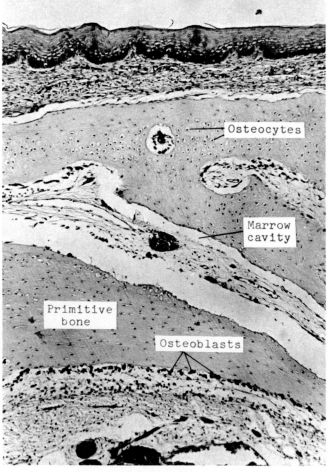

Fig. 13c   Intramembranous bone formation in the skull (later) x.s. x40.

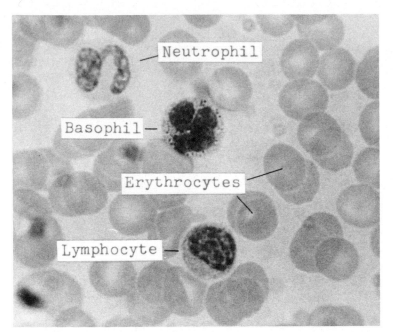

Fig. 14a   Human blood w.m. x1000. The cytoplasm of the neutrophil stains only faintly and is difficult to see in this photo.

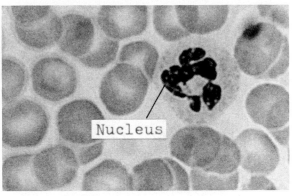

Fig. 14b   Neutrophil w.m. x1000. The nucleus has many lobes and is polymorphic in these cells. Cytoplasmic granules are faintly visible.

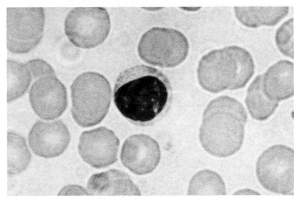

Fig. 14c   Lymphocyte w.m. x1000. The nucleus usually takes up 80-90% of the cell.

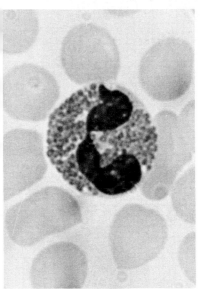

Fig. 14d   Eosinophil w.m. x1000. The granules in the cytoplasm will stain red.

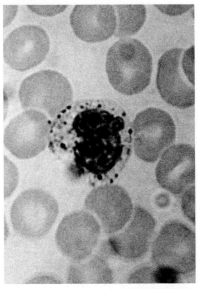

Fig. 14e   Basophil w.m. x1000. The dark granules in the cytoplasm will stain a dark blue or purple.

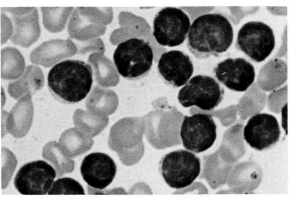

Fig. 14f   Chronic lymphocytic leukemia. The patient died 4 days after this sample was taken. Notice the abnormally high number of lymphocytes.

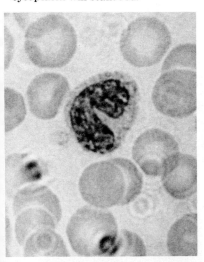

Fig. 14g   Monocyte w.m. x1000. Monocytes are often up to 2x the size of other WBC's and commonly have a horse-shoe shaped nucleus.

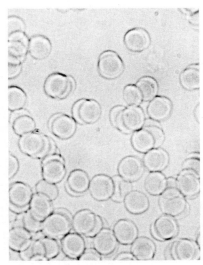

Fig. 14h   Erythrocytes (live) w.m. x430. The biconcave shape is easily seen.

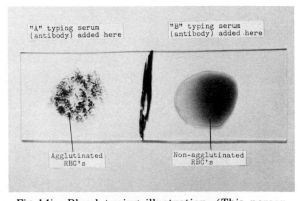

Fig. 14i   Blood typing illustration. (This person had type "A" blood.) "A" antibody (typing serum) was added to the drop of blood on the left side and "B" antibody was added to the right. Agglutination occurred on the left side between the "A" antibodies and the "A" antigens on the person's RBC's.

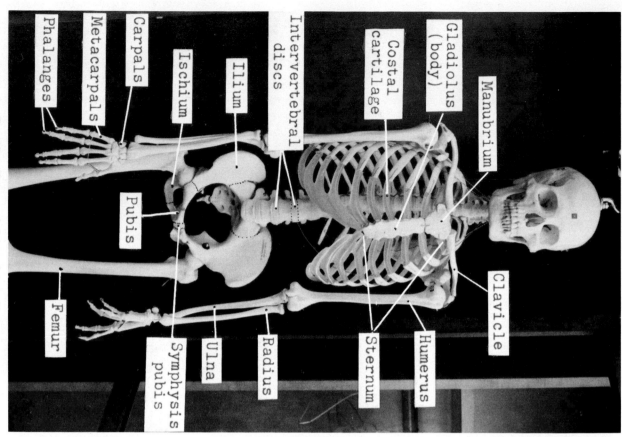

Fig. 15a Human skeleton, front view.

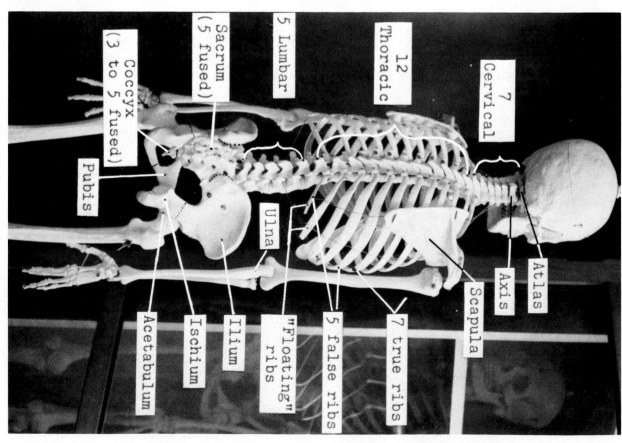

Fig. 15b Human skeleton, rear view.

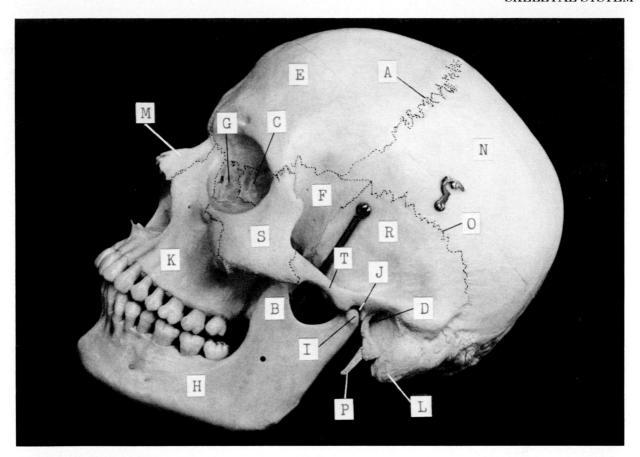

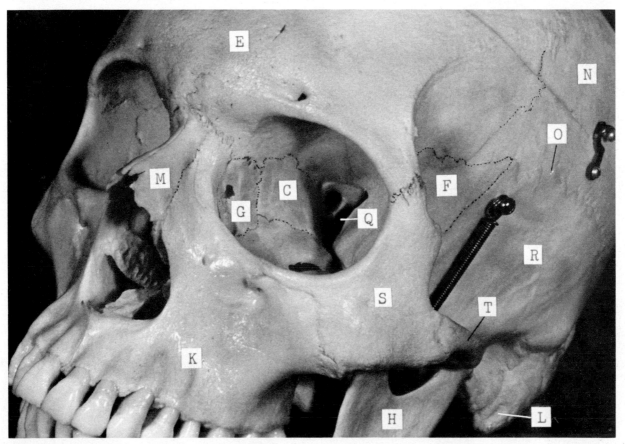

Fig. 16a & 16b   Human skull.

A - Coronal suture
B - Coronoid process
C - Ethmoid
D - External acoustic meatus
E - Frontal
F - Sphenoid
G - Lacrimal

H - Mandible
I - Mandibular condyle
J - Mandibular (Glenoid) fossa
K - Maxilla
L - Mastoid process
M - Nasal
N - Parietal

O - Squamosal suture
P - Styloid process
Q - Superior orbital fissure
R - Temporal
S - Zygomatic (Malar)
T - Zygomatic process of temporal

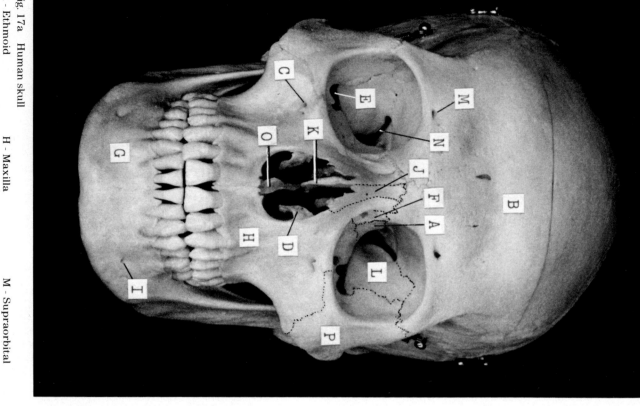

Fig. 17a   Human skull

A - Ethmoid
B - Frontal
C - Infraorbital foramen
D - Inferior nasal concha
E - Inferior orbital fissure
F - Lacrimal
G - Mandible

H - Maxilla
I - Mental foramen
J - Nasal
K - Perpendicular plate
    of Ethmoid
L - Orbital surface of
    Sphenoid

M - Supraorbital
    foramen
N - Superior orbital
    fissure
O - Vomer
P - Zygomatic (malar)

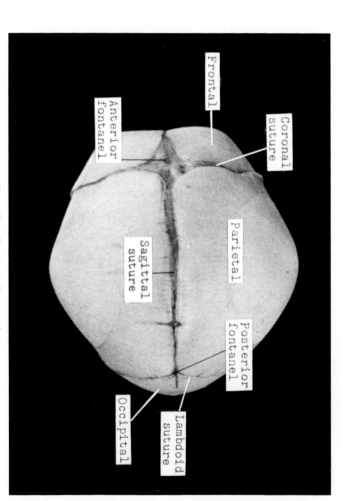

Fig. 17c   Fetal skull, top view.

Anterior fontanel

Frontal

Coronal suture

Sagittal suture

Parietal

Occipital

Lambdoid suture

Posterior fontanel

Fig. 17b   Fetal skull, side view.

Anterolateral (sphenoidal) fontanel

Sphenoid

Frontal

Coronal suture

Parietal

Squamosal suture

Temporal

Posterolateral (mastoid) fontanel

Occipital

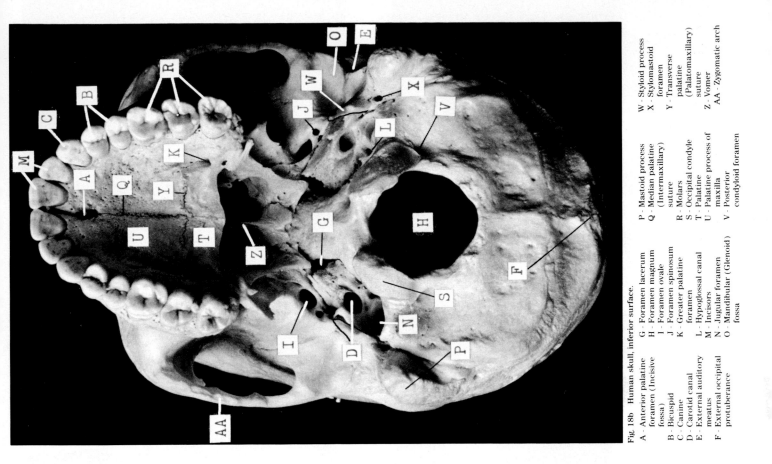

Fig. 18b Human skull, inferior surface.

A - Anterior palatine foramen (Incisive fossa)
B - Bicuspid
C - Canine
D - Carotid canal
E - External auditory meatus
F - External occipital protuberance
G - Foramen lacerum
H - Foramen magnum
I - Foramen ovale
J - Foramen spinosum
K - Greater palatine foramen
L - Hypoglossal canal
M - Incisors
N - Jugular foramen
O - Mandibular (Glenoid) fossa
P - Mastoid process
Q - Median palatine (Intermaxillary) suture
R - Molars
S - Occipital condyle
T - Palatine
U - Palatine process of maxilla
V - Posterior condyloid foramen
W - Styloid process
X - Stylomastoid foramen
Y - Transverse palatine (Palatomaxillary) suture
Z - Vomer
AA - Zygomatic arch

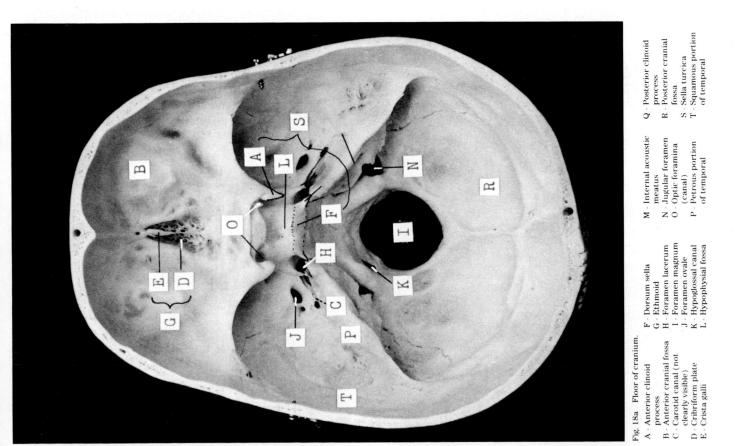

Fig. 18a Floor of cranium.

A - Anterior clinoid process
B - Anterior cranial fossa
C - Carotid canal (not clearly visible)
D - Cribriform plate
E - Crista galli
F - Dorsum sella
G - Ethmoid
H - Foramen lacerum
I - Foramen magnum
J - Foramen ovale
K - Hypoglossal canal
L - Hypophysial fossa
M - Internal acoustic meatus
N - Jugular foramen
O - Optic foramina (canal)
P - Petrous portion of temporal
Q - Posterior clinoid process
R - Posterior cranial fossa
S - Sella turcica
T - Squamous portion of temporal

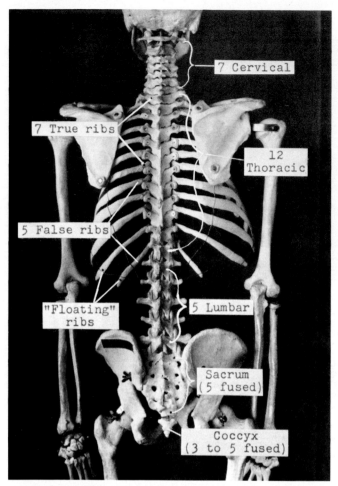

Fig. 19a Vertebral column, dorsal view.

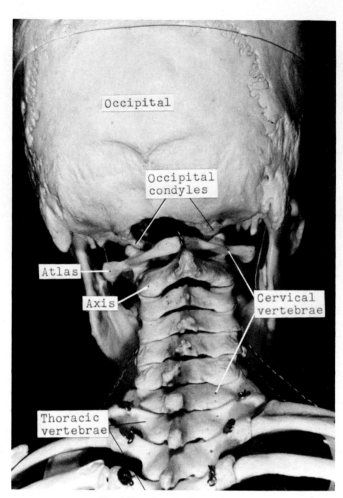

Fig. 19b Cervical vertebrae.

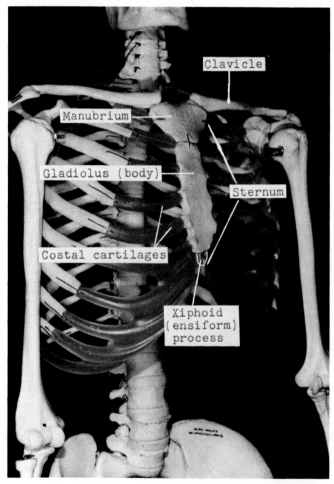

Fig. 19c Thorax, ventral view.

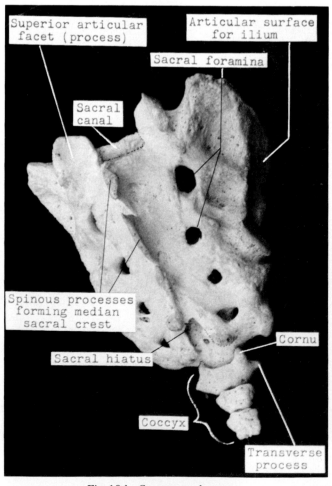

Fig. 19d Sacrum and coccyx.

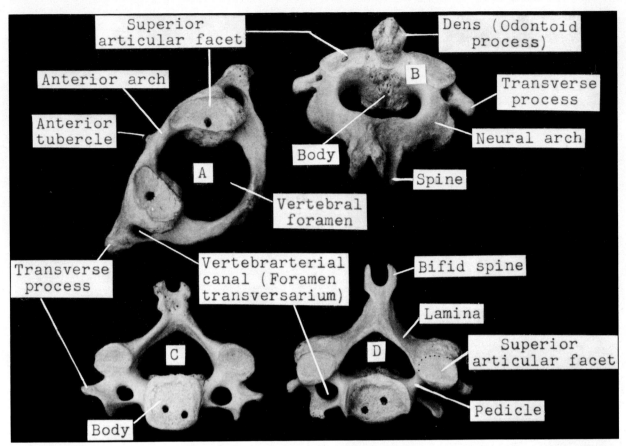

Fig. 20a   Cervical vertebrae: A-Atlas, B-Axis, C-typical cervical vertebra, inferior surface, D-typical cervical vertebra, superior surface. (The two small holes in the body held wires in the articulated skeleton.)

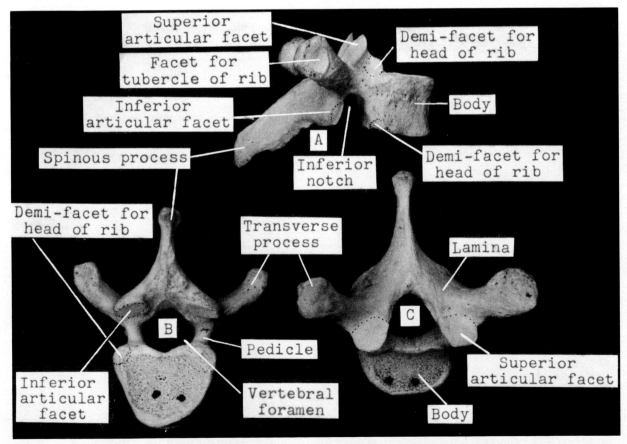

Fig. 20b   Thoracic vertebrae: A-lateral surface, B-inferior surface, C-superior surface.

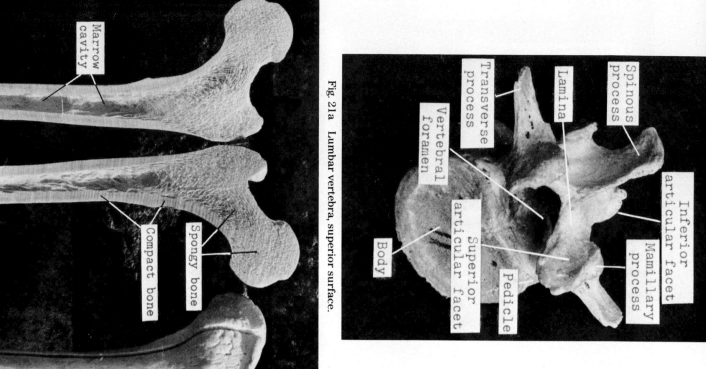

Fig. 21a Lumbar vertebra, superior surface.

Transverse process
Vertebral foramen
Spinous process
Lamina
Inferior articular facet
Mamillary process
Superior articular facet
Pedicle
Body

Fig. 21c Sectioned human femur.

Marrow cavity
Compact bone
Spongy bone
Nutrient foramina

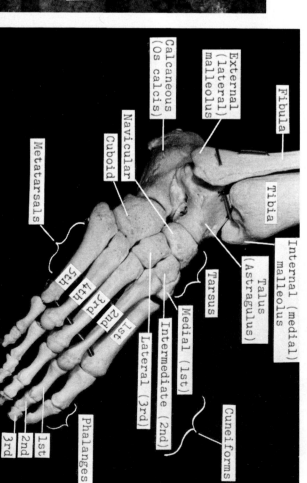

Fig. 21d Right foot, dorsal aspect.

Metatarsals
Calcaneous (Os calcis)
External (lateral) malleolus
Fibula
Navicular
Cuboid
Tibia
Internal (medial) malleolus
Talus (Astragulus)
Tarsus
Medial (1st)
Intermediate (2nd)
Lateral (3rd)
Cuneiforms
Phalanges
5th
4th
3rd
2nd
1st
3rd
2nd
1st

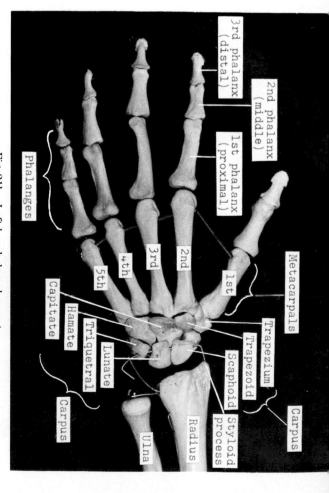

Fig. 21b Left hand, dorsal aspect.

3rd phalanx (distal)
2nd phalanx (middle)
1st phalanx (proximal)
Phalanges
Metacarpals
5th
4th
3rd
2nd
1st
Capitate
Hamate
Triquetral
Lunate
Carpus
Ulna
Radius
Styloid process
Scaphoid
Trapezium
Trapezoid
Carpus

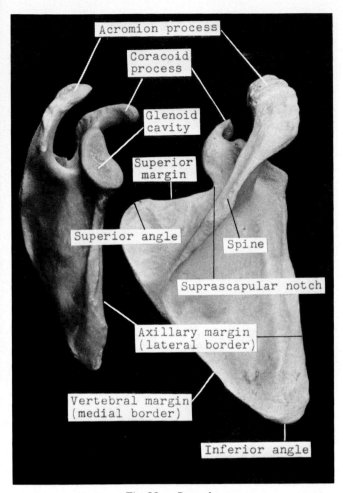

Fig. 22a Scapula.

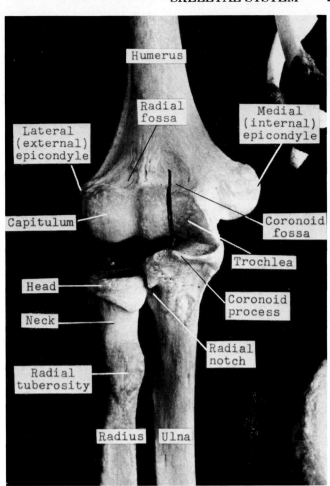

Fig. 22b Right elbow region, articulated, anterior view.

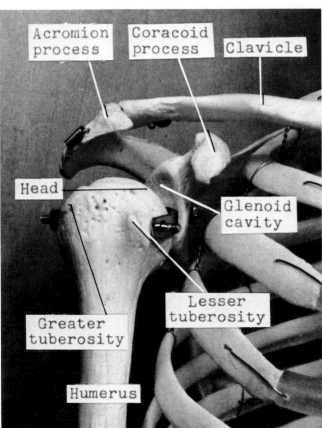

Fig. 22c Right shoulder region.

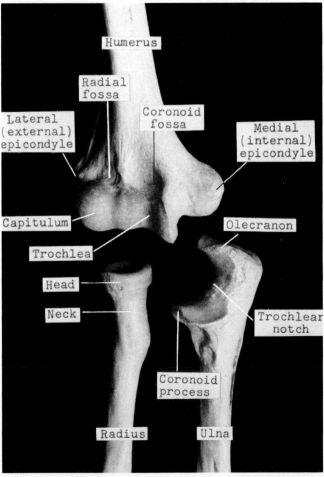

Fig. 22d Right elbow region, disarticulated.

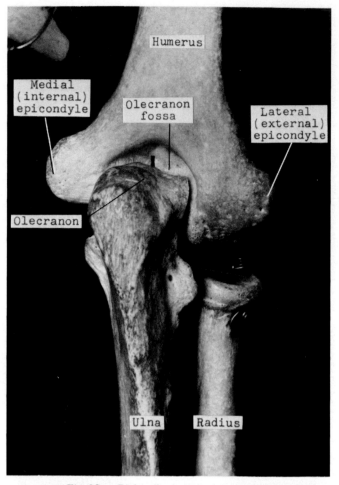

Fig. 23a   Right elbow region, posterior.

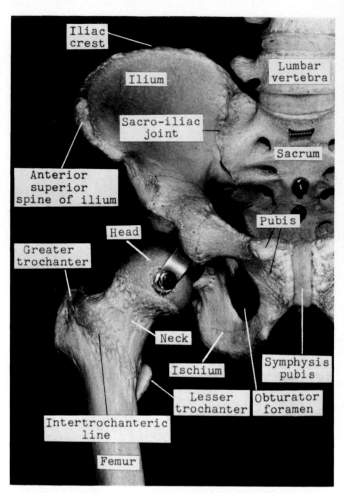

Fig. 23b   Pelvis, right hip region, anterior.

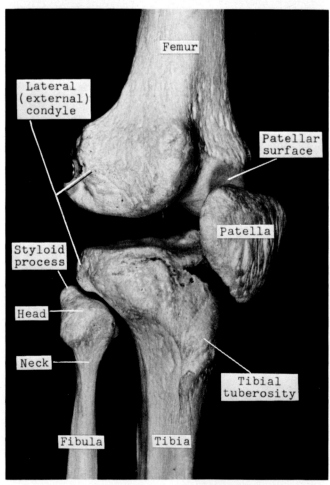

Fig. 23c   Right knee region, anterior.

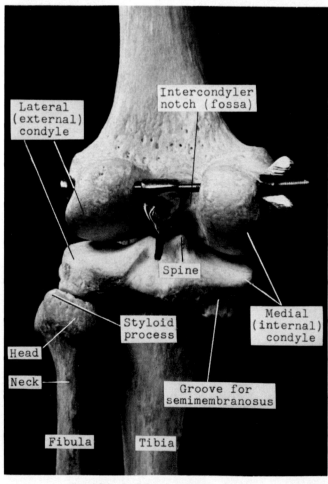

Fig. 23d   Left knee region, posterior.

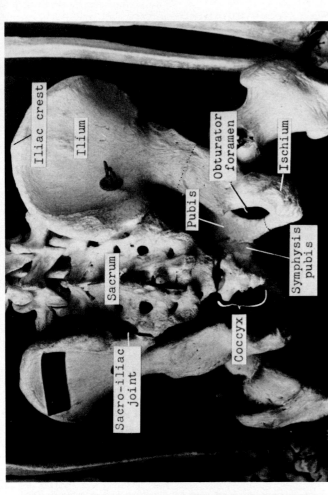

Fig. 24a   Male pelvis, anterior view. (Coccyx missing on this specimen.)

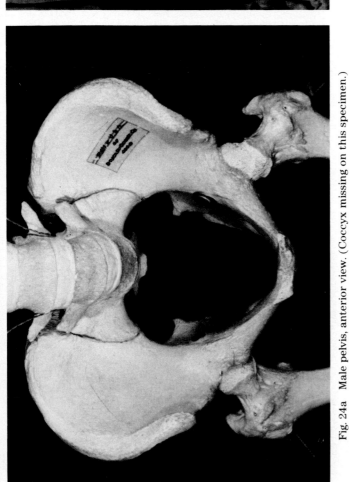

Fig. 24c   Female pelvis, anterior view. Notice the flared crest of the ilium, and the larger opening in the center of the pelvis.

Iliac crest

Ilium

Obturator foramen

Pubis

Ischium

Sacrum

Symphysis pubis

Coccyx

Sacro-iliac joint

Fig. 24b   Pelvis, posterior view.

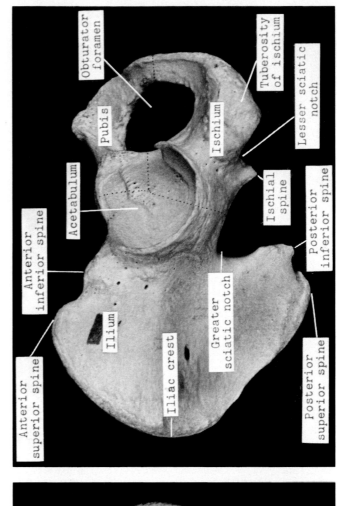

Obturator foramen

Tuberosity of ischium

Pubis

Ischium

Lesser sciatic notch

Acetabulum

Anterior inferior spine

Ischial spine

Posterior inferior spine

Anterior superior spine

Ilium

Iliac crest

Greater sciatic notch

Posterior superior spine

Fig. 24d   Os Coxa.

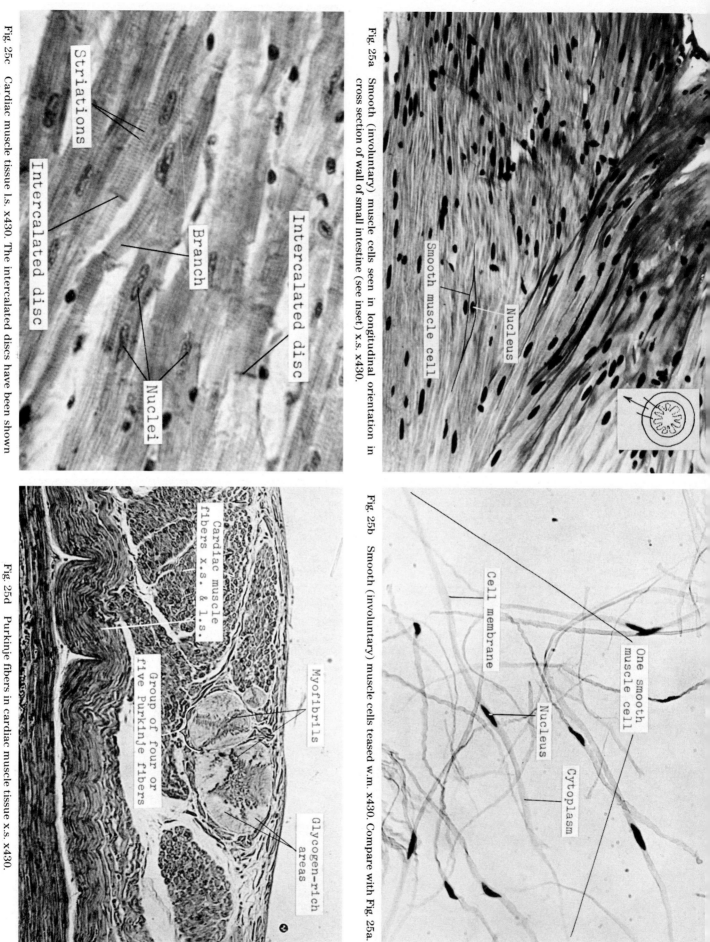

Fig. 25a   Smooth (involuntary) muscle cells seen in longitudinal orientation in cross section of wall of small intestine (see inset) x.s. x430.

Smooth muscle cell

Nucleus

Fig. 25c   Cardiac muscle tissue l.s. x430. The intercalated discs have been shown with the electron microscope to be the abutting cell membranes at the ends of two adjacent cells.

Striations

Intercalated disc

Branch

Nuclei

Intercalated disc

Fig. 25b   Smooth (involuntary) muscle cells teased w.m. x430. Compare with Fig. 25a.

Cell membrane

Nucleus

One smooth muscle cell

Cytoplasm

Fig. 25d   Purkinje fibers in cardiac muscle tissue x.s. x430.

Cardiac muscle fibers x.s. & l.s.

Group of four or five Purkinje fibers

Myofibrils

Glycogen-rich areas

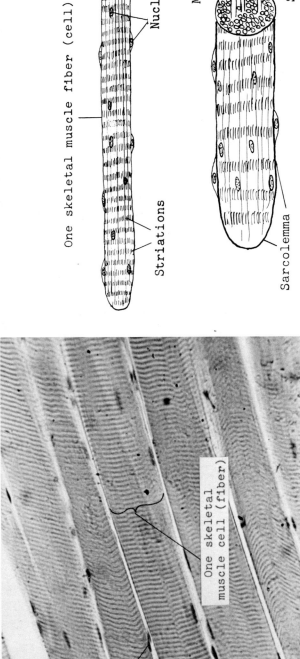

One skeletal muscle fiber (cell)

Nuclei

Striations

Myofibrils

Striations

Sarcolemma

Fig. 26b   Diagrams of skeletal muscle cells. The lower cell has been cut in cross section to show features of the cells interior.

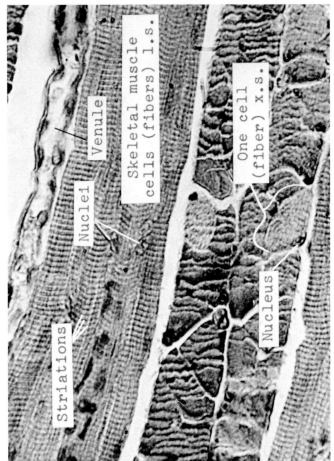

Venule

Nuclei

Skeletal muscle cells (fibers) l.s.

One cell (fiber) x.s.

Striations

Nucleus

Fig. 26d   Skeletal muscle fibers from the tongue in x.s. and l.s. x430.

One skeletal muscle cell (fiber)

Nuclei

Striations

Fig. 26a   Skeletal (voluntary) (striated) muscle fibers (cells) l.s. x430.

Striations

Myofibrils

Muscle fiber (one cell)

Muscle fiber (one cell)

Fig. 26c   Exposed ends of torn frog skeletal muscle fibers showing myofibrils and repeated banding pattern of striations w.m. x430. Compare with Fig. 26b.

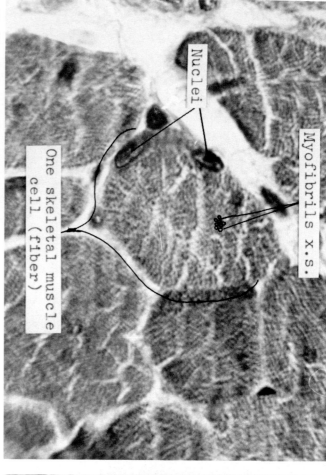

Fig. 27a   Skeletal muscle fibers (cells) x.s. x430. Notice how closely each skeletal muscle cell resembles a coaxial cable of myofibrils. Compare with Fig. 26b.

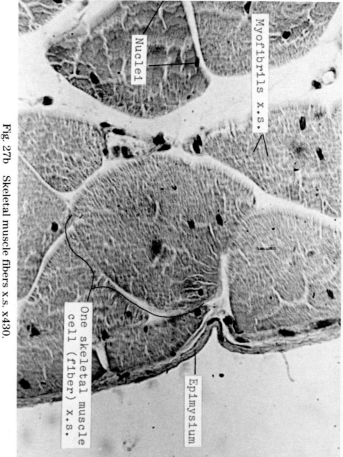

Fig. 27b   Skeletal muscle fibers x.s. x430.

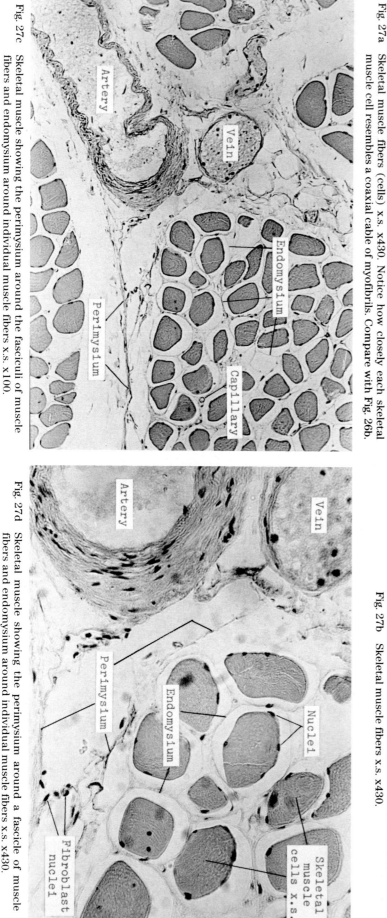

Fig. 27c   Skeletal muscle showing the perimysium around the fasciculi of muscle fibers and endomysium around individual muscle fibers x.s. x100.

Fig. 27d   Skeletal muscle showing the perimysium around a fascicle of muscle fibers and endomysium around individual muscle fibers x.s. x430.

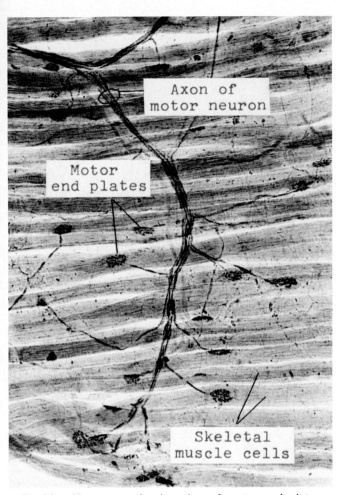

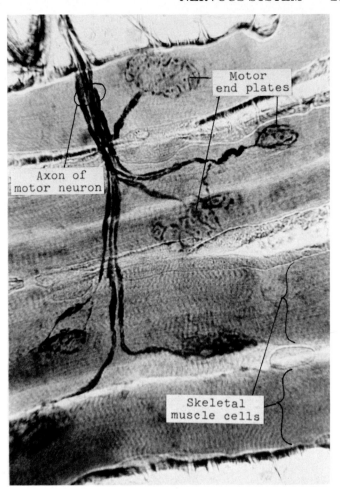

Fig. 28a   Neuromuscular junction of motor end plates and skeletal muscle cells w.m. x100.

Fig. 28b   Neuromuscular junction of motor end plates and skeletal muscle cells w.m. x430. Individual motor end plates each supply one skeletal muscle fiber.

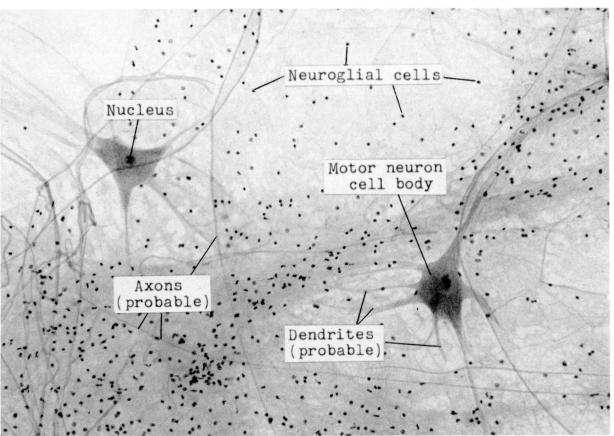

Fig. 28c   Neurons (nerve cells) from the spinal cord of an ox w.m. x100. Since neurons in the spinal cord are either motor or association neurons, the long cell processes would probably be axons and the short ones dendrites. However, it is difficult to be sure from a photo such as this.

Know this one
↓

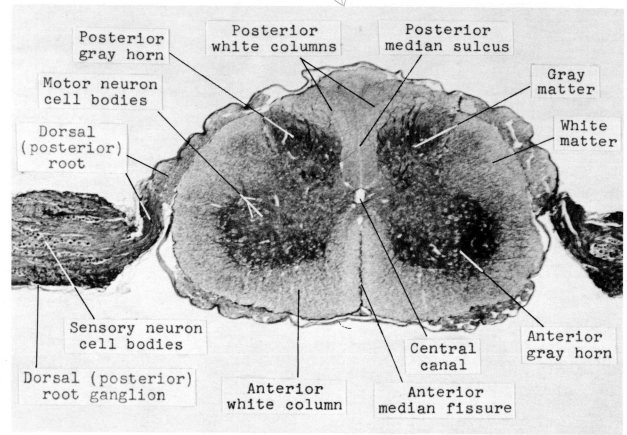

Posterior
gray horn

Posterior
white columns

Posterior
median sulcus

Gray
matter

Motor neuron
cell bodies

White
matter

Dorsal
(posterior)
root

Sensory neuron
cell bodies

Dorsal (posterior)
root ganglion

Anterior
white column

Central
canal

Anterior
median fissure

Anterior
gray horn

Fig. 29a   Spinal cord x.s. x20.

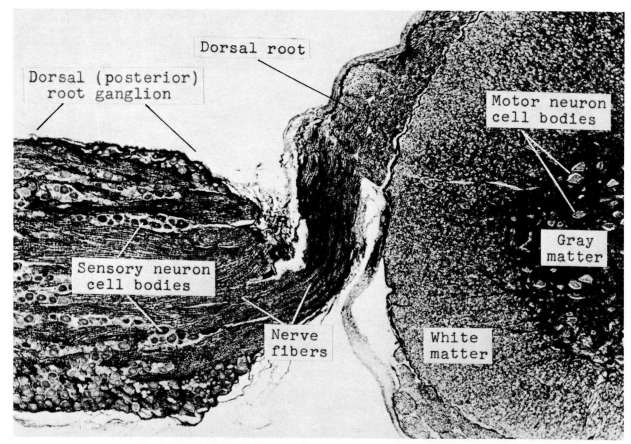

Dorsal root

Dorsal (posterior)
root ganglion

Motor neuron
cell bodies

Gray
matter

Sensory neuron
cell bodies

Nerve
fibers

White
matter

Fig. 29b   Spinal cord and dorsal root ganglion x.s. x40.

multipolar *know this* ↓

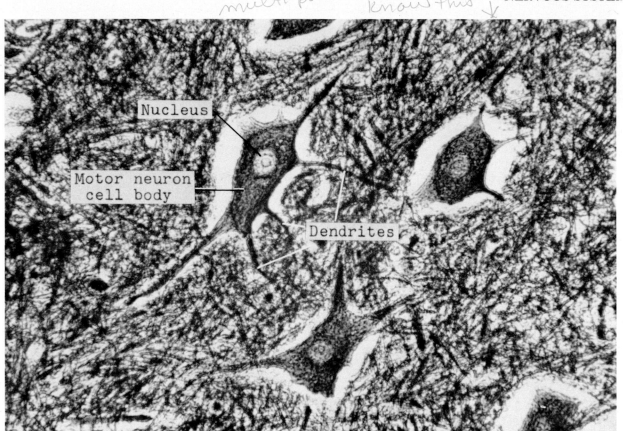

Nucleus

Motor neuron cell body

Dendrites

Fig. 30a   Motor neurons in gray matter of spinal cord x.s. x430. (See Fig. 29b.)

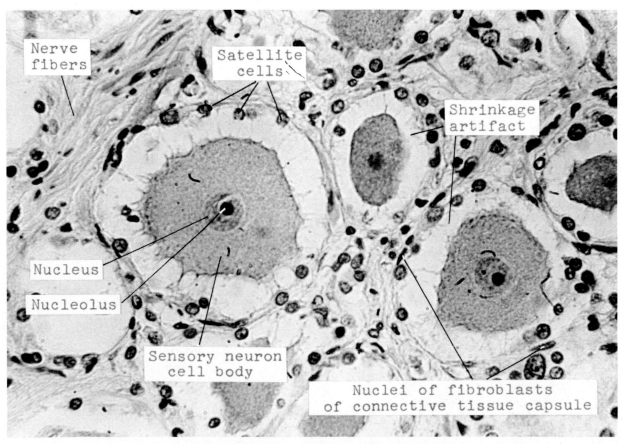

Nerve fibers

Satellite cells

Shrinkage artifact

Nucleus

Nucleolus

Sensory neuron cell body

Nuclei of fibroblasts of connective tissue capsule

Fig. 30b   Dorsal root ganglion x.s. x430. (See Fig. 29b.)

Brown yellow orange
yellow balls
brown fibers

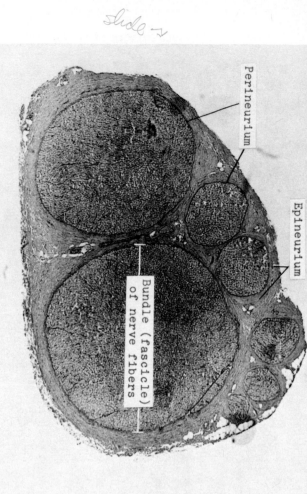

Fig. 31a Peripheral nerve x.s. x40.

Perineurium

Epineurium

Bundle (fascicle) of nerve fibers

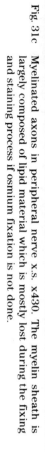

Fig. 31c Myelinated axons in peripheral nerve x.s. x430. The myelin sheath is largely composed of lipid material which is mostly lost during the fixing and staining process if osmium fixation is not done.

Epineurium

Myelin sheath (remnant)

Neurilemma

Axon

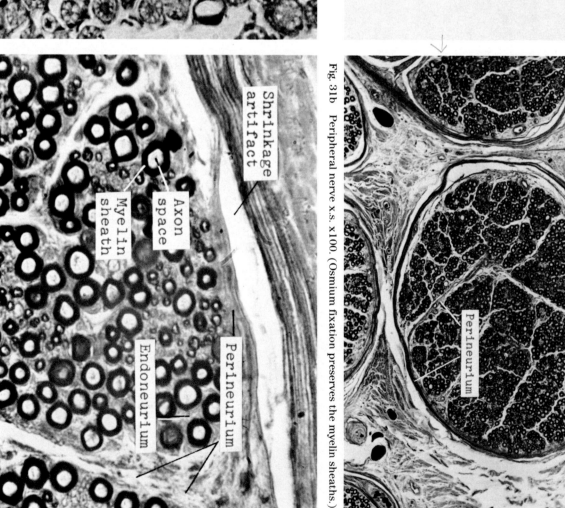

Fig. 31b Peripheral nerve x.s. x100. (Osmium fixation preserves the myelin sheaths.)

Epineurium

Perineurium

Fig. 31d Myelinated axons in peripheral nerve x.s. x430. (Osmium fixation preserves the myelin sheaths.)

Shrinkage artifact

Myelin sheath

Axon space

Endoneurium

Perineurium

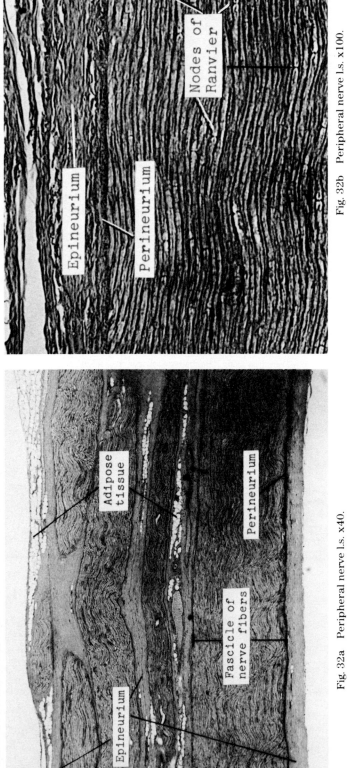

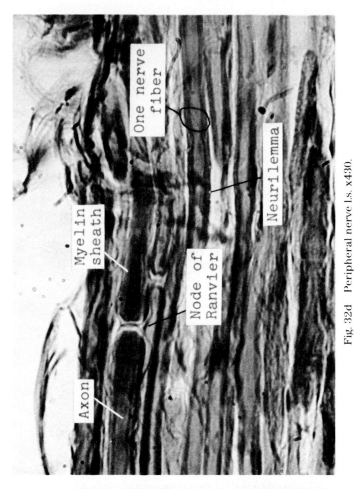

Fig. 32a Peripheral nerve l.s. x40.

Fig. 32b Peripheral nerve l.s. x100.

Fig. 32c Peripheral nerve l.s. x430. The node of Ranvier is a short interval where the axon is not covered by a myelin sheath.

Fig. 32d Peripheral nerve l.s. x430.

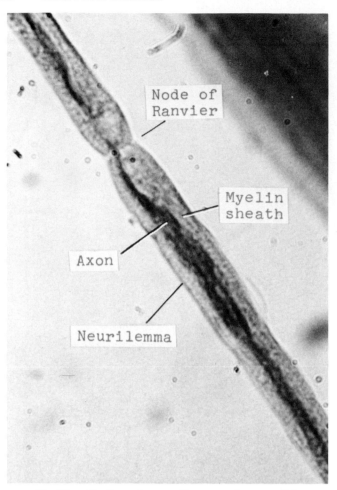

Fig. 33a   Myelinated nerve fiber, teased w.m. x430.

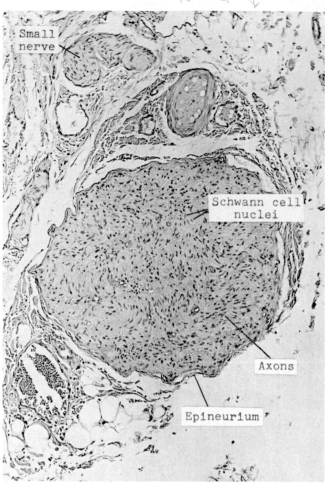

Fig. 33b   Peripheral nerve x.s. x100.

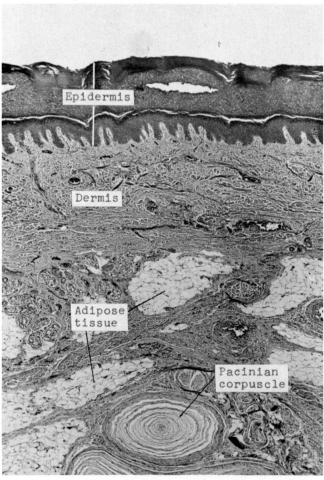

Fig. 33c   Pacinian corpuscle in thick skin x.s. x40.
Pacinian corpuscles are also present in viscera.

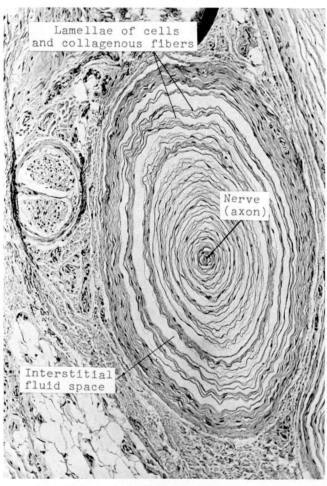

Fig. 33d   Pacinian corpuscle x.s. x100. This receptor is
sensitive to pressure.

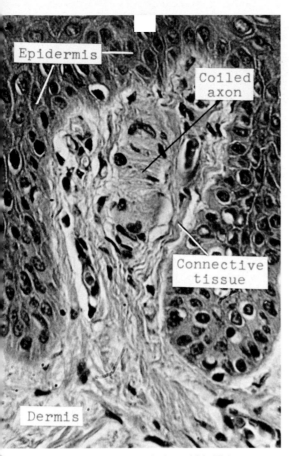

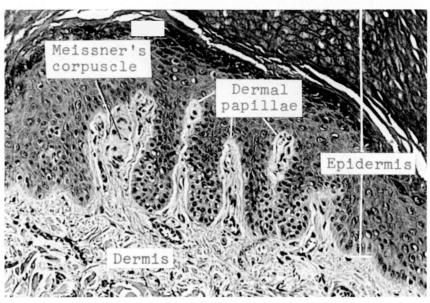

Fig. 34b   Meissner's corpuscle l.s. x430. (Also see Fig. 47c.)

Fig. 34a   Meissner's corpuscle l.s. x100. This receptor responds to tactile stimuli and is located just below the epidermis.

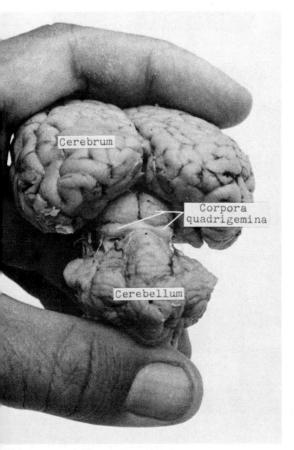

Fig. 34c   Sheep brain, bent to expose the midbrain.

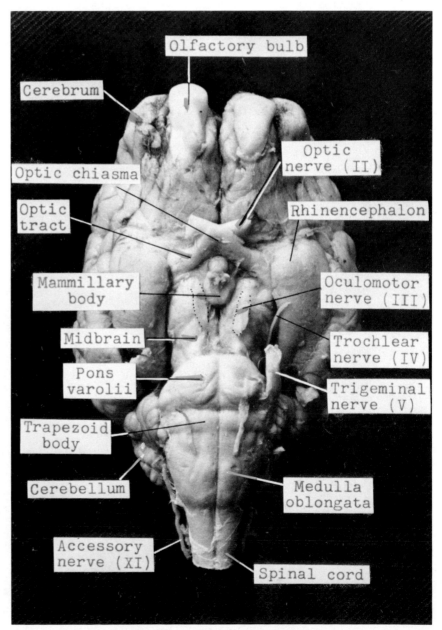

Fig. 34d   Sheep brain, ventral view.

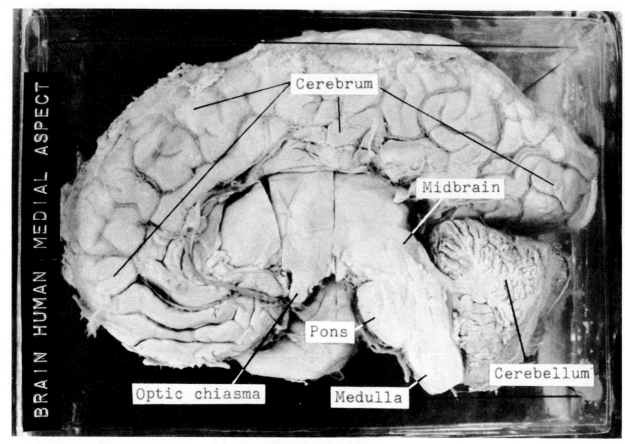

Fig. 35a   Human brain l.s.

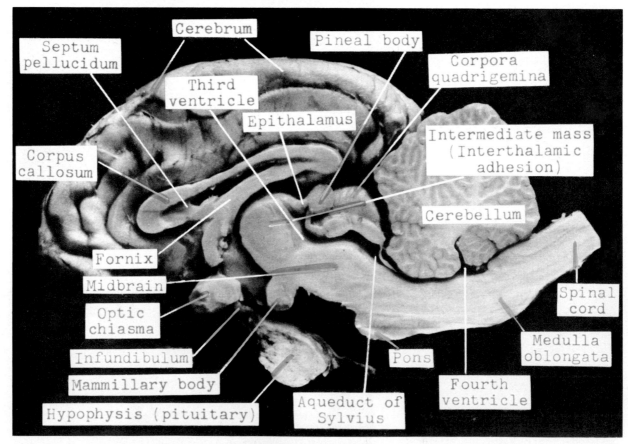

Fig. 35b   Sheep brain l.s.

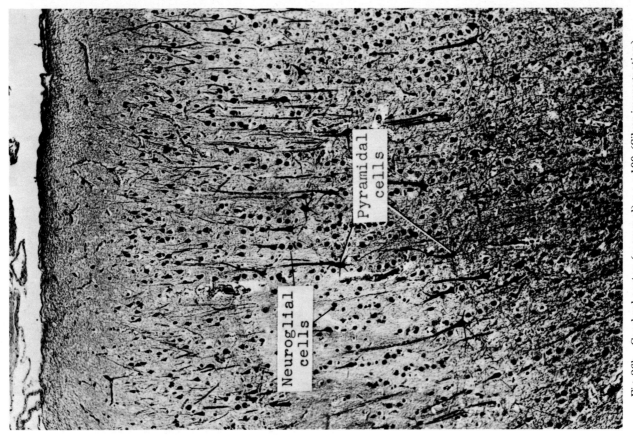

Pyramidal cells

Neuroglial cells

Fig. 36b   Cerebral cortex (mammal) x.s. x100. (Silver impregnation)

Purple & white spots

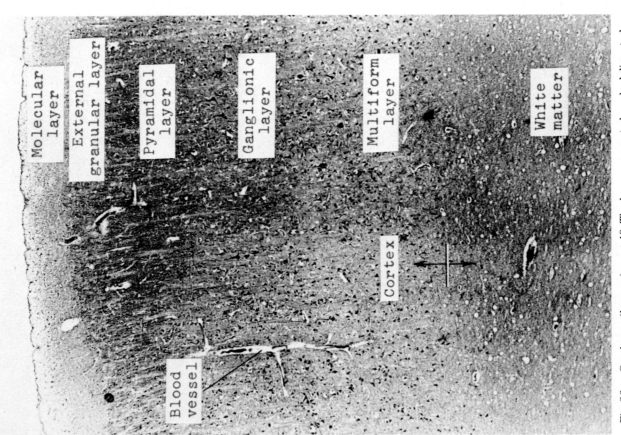

Molecular layer

External granular layer

Pyramidal layer

Ganglionic layer

Multiform layer

White matter

Cortex

Blood vessel

Fig. 36a   Cerebrum (human) x.s. x40. The layers are not sharply delineated. (Silver impregnation)

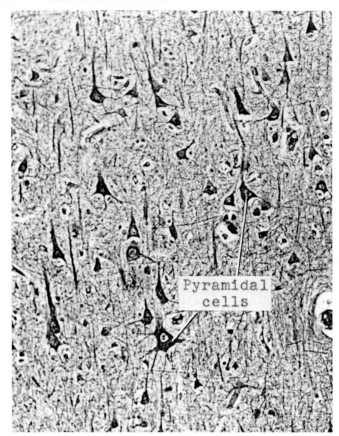

Fig. 37a   Cerebral cortex (human) x.s. x100.

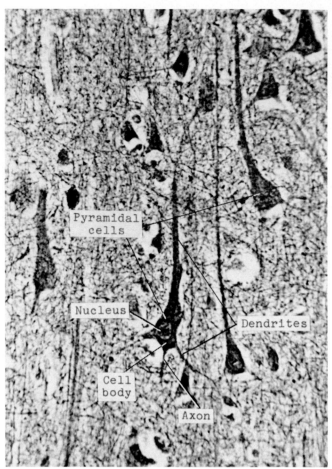

Fig. 37b   Pyramidal neurons in cerebral cortex (human) x.s. x430.

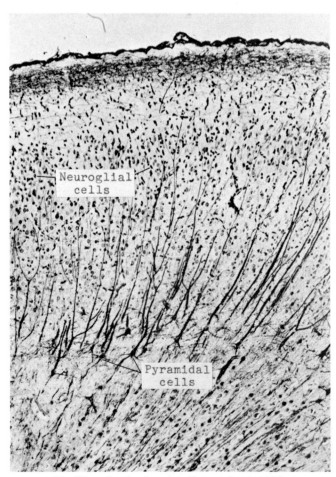

Fig. 37c   Cerebral cortex (human) x.s. x100.

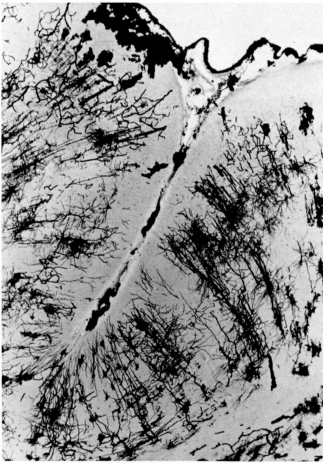

Fig. 37d   Pyramidal neurons in cerebral cortex x.s. x40. Notice the many branches these neurons have. (Golgi stain)

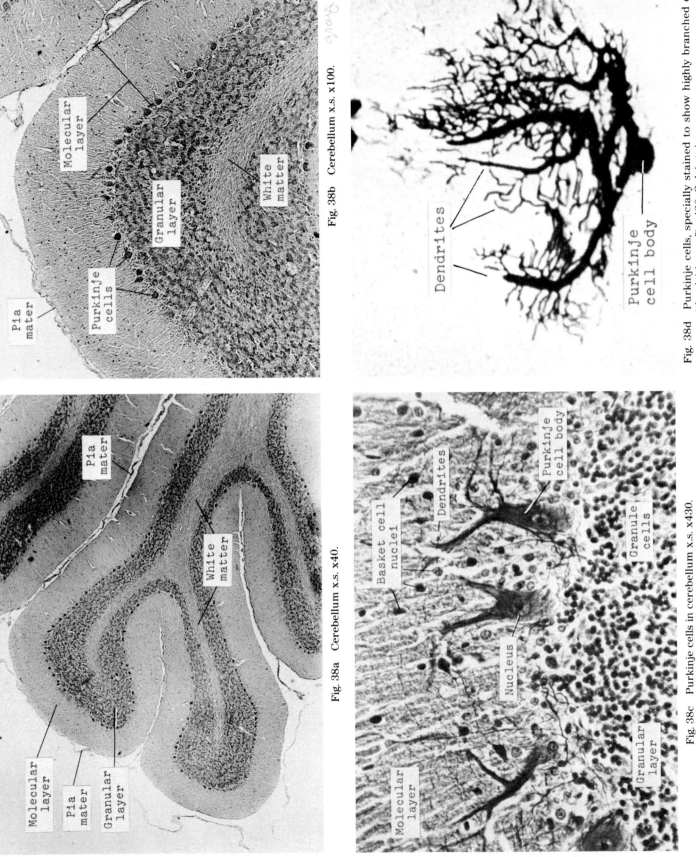

Fig. 38b Cerebellum x.s. x100.

graag bug spots

Fig. 38d Purkinje cells, specially stained to show highly branched dendrites (the "dendritic tree") x430. Golgi stain.

Fig. 38a Cerebellum x.s. x40.

Fig. 38c Purkinje cells in cerebellum x.s. x430.

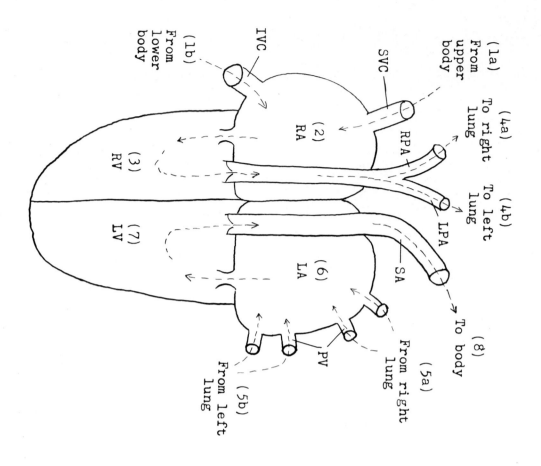

Fig. 39a   Diagram of human heart circulation. (Use the numbers and follow the path of circulation through the heart.)

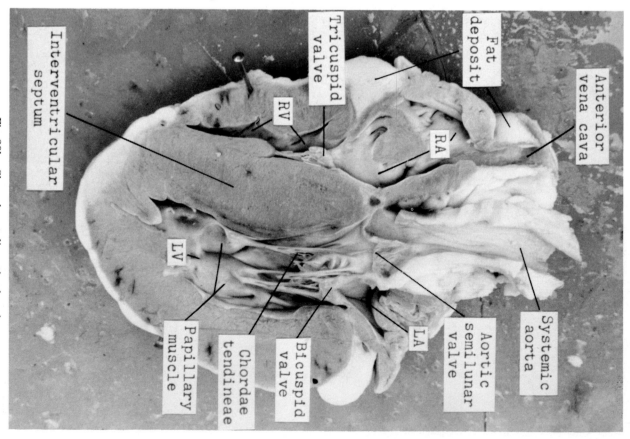

Fig. 39b   Sheep heart dissection l.s. x1.

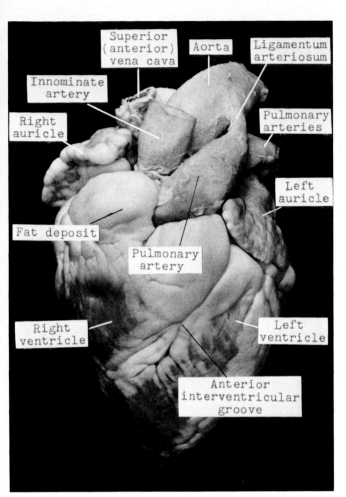

Fig. 40a   Sheep heart, ventral view.

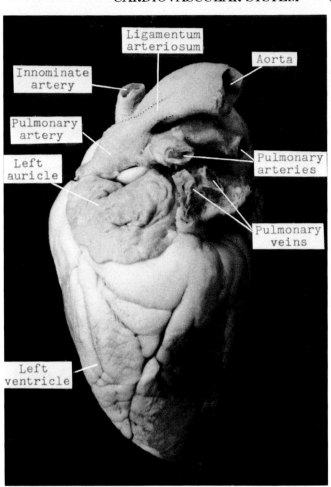

Fig. 40b   Sheep heart, dorso-lateral view.

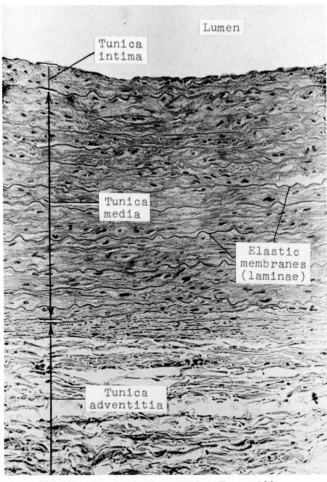

Fig. 40c   Aorta (elastic artery) wall x.s. x100.

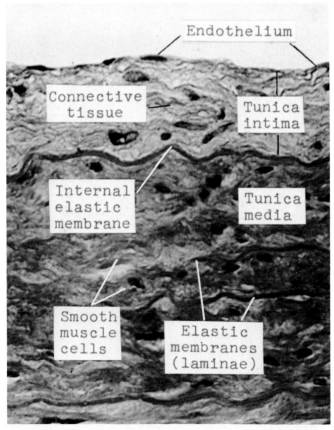

Fig. 40d   Aorta (elastic artery) wall x.s. x430. The elasticity enables the artery to withstand the higher blood pressures coming directly from the ventricle. The endothelium is simple squamous epithelium.

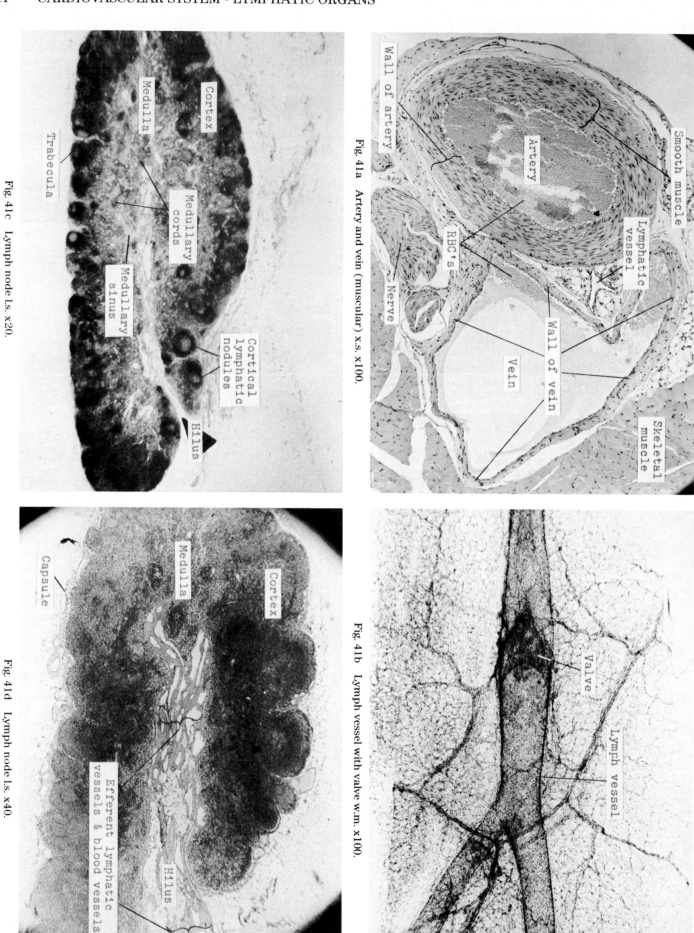

Fig. 41a   Artery and vein (muscular) x.s. x100.

Fig. 41b   Lymph vessel with valve w.m. x100.

Fig. 41c   Lymph node l.s. x20.

Fig. 41d   Lymph node l.s. x40.

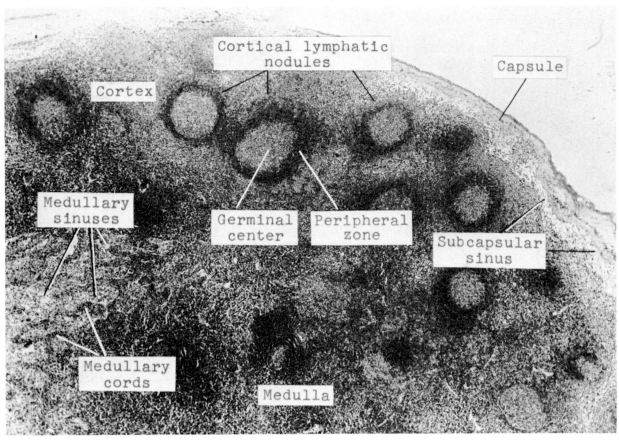

Fig. 42a   Lymph node x.s. x40.

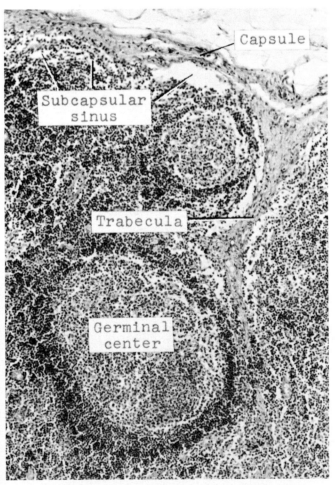

Fig. 42b   Lymph node, cortex x.s. x100.

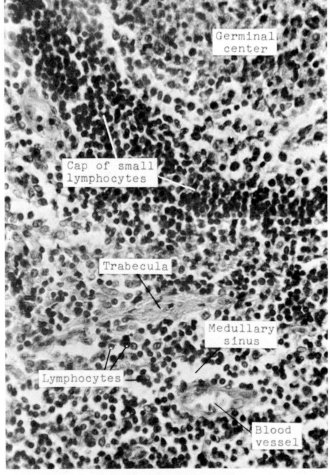

Fig. 42c   Lymph node, detail of cortical nodule and peripheral zone x.s. x430.

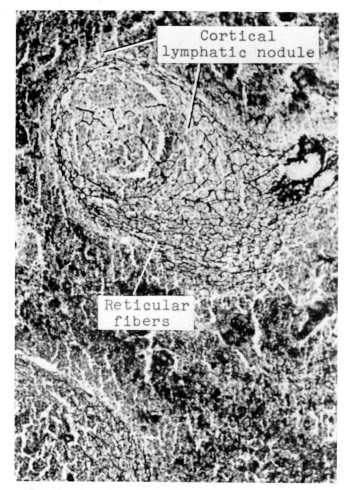

Fig. 43a   Lymph node showing reticular fibers x.s. x100.

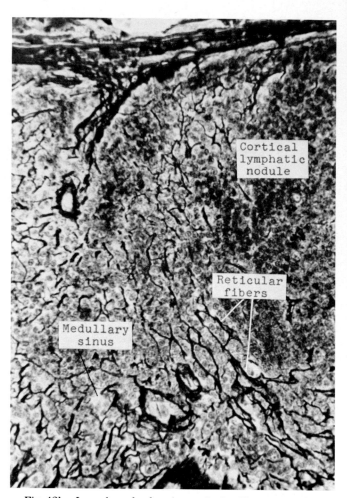

Fig. 43b   Lymph node showing reticular fibers x.s. x430.

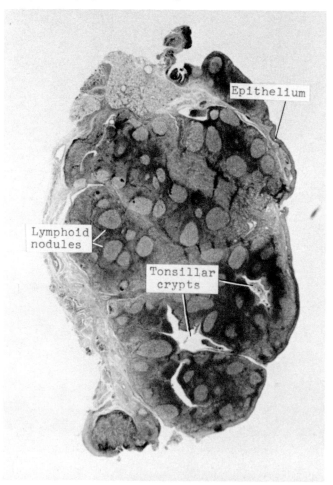

Fig. 43c   Palatine tonsil (human) x.s. x20.

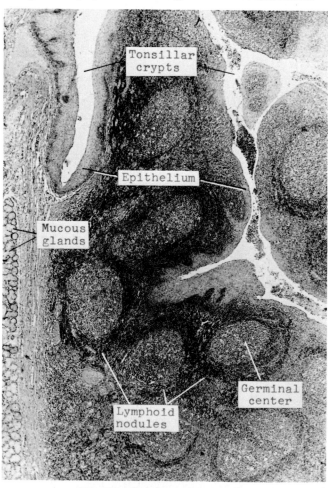

Fig. 43d   Palatine tonsil x.s. x40.

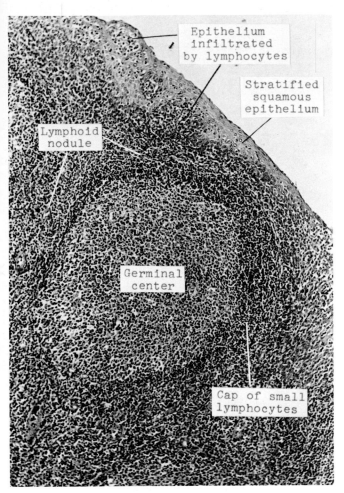

Epithelium
infiltrated
by lymphocytes

Stratified
squamous
epithelium

Lymphoid
nodule

Germinal
center

Cap of small
lymphocytes

Fig. 44a   Palatine tonsil x.s. x100.

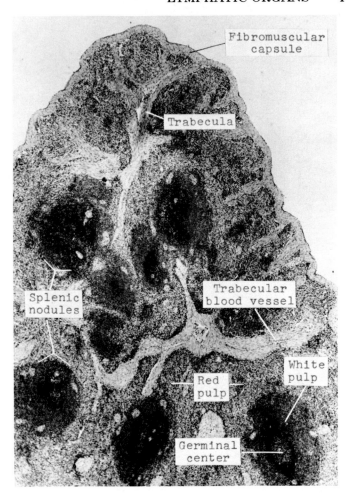

Fibromuscular
capsule

Trabecula

Splenic
nodules

Trabecular
blood vessel

Red
pulp

White
pulp

Germinal
center

Fig. 44b   Spleen (dog) x.s. x40.

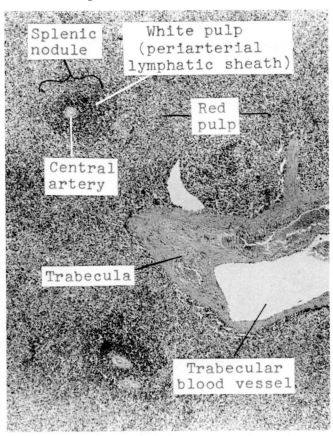

Splenic
nodule

White pulp
(periarterial
lymphatic sheath)

Red
pulp

Central
artery

Trabecula

Trabecular
blood vessel

Fig. 44c   Spleen (human) x.s. x40.

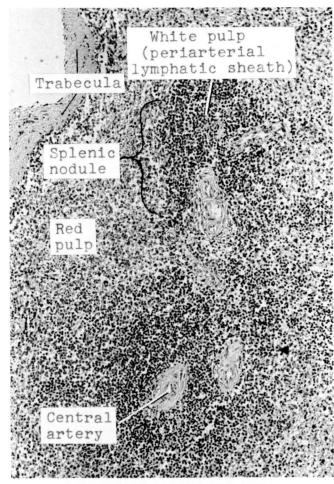

Trabecula

White pulp
(periarterial
lymphatic sheath)

Splenic
nodule

Red
pulp

Central
artery

Fig. 44d   Spleen (human) x.s. x100.

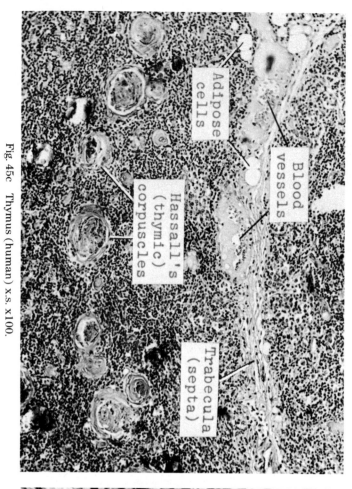

Fig. 45c Thymus (human) x.s. x100.

Fig. 45a Thymus (cat) x.s. x100.

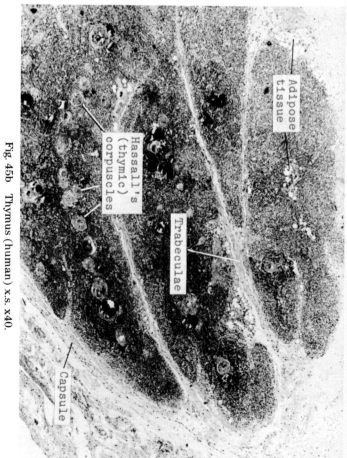

Fig. 45b Thymus (human) x.s. x40.

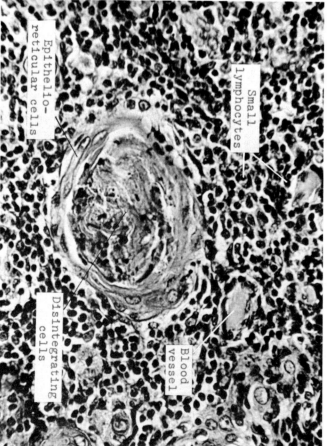

Fig. 45d Hassall's corpuscle x.s. x430. Hassall's corpuscles are characteristic features of the thymus, increasing in number with age, however their function is not known.

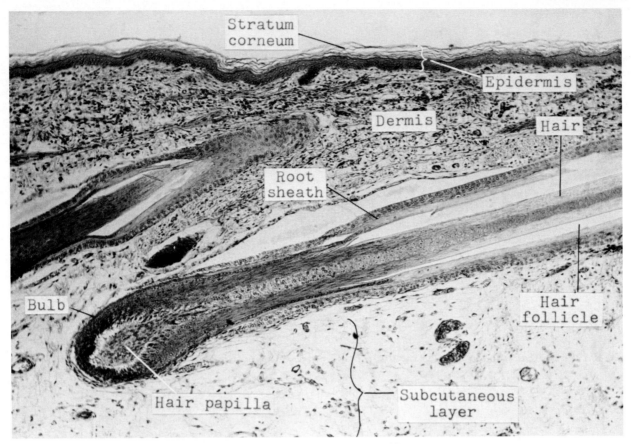

Fig. 46a   Skin and hair l.s. x100.

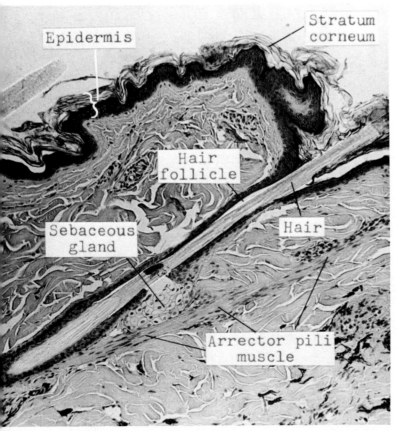

Fig. 46b   Skin and hair l.s. x100. The Arrector pili muscle "erects" the hair in animals for insulation against cold and/or to appear larger during fights. In humans, the erected hair produces a fold of skin in front of it - the "goose bump."

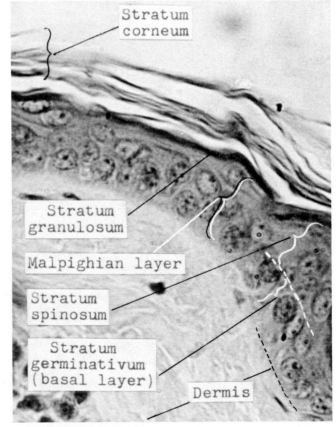

Fig. 46c   Thin skin, layers of the epidermis x.s. x430.

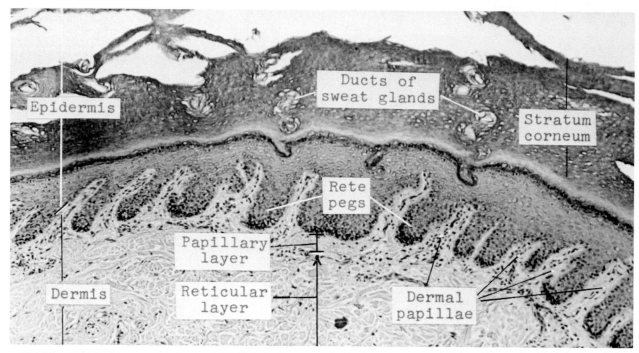

Fig. 47a   Thick skin (monkey) x.s. x100. This is the type of skin found on the palms of the hands and soles of the feet. (Also see Fig. 33c.)

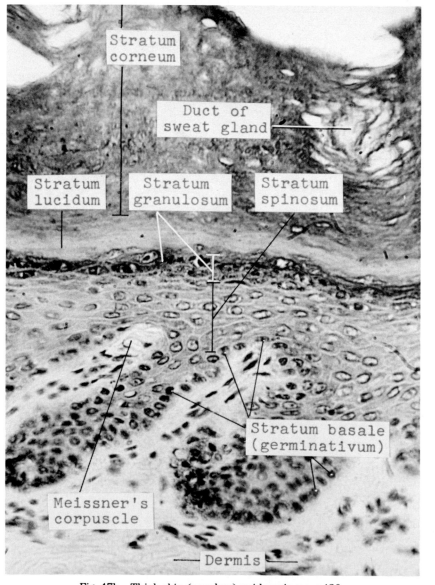

Fig. 47b   Thick skin (monkey) epidermis x.s. x430.

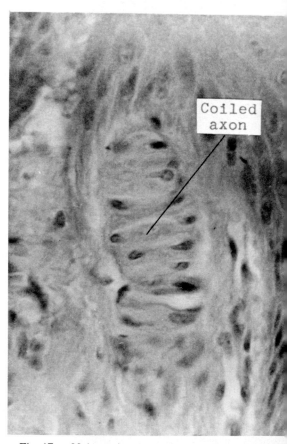

Fig. 47c   Meissner's corpuscle l.s. x430. These are tactile (touch) receptors. (Also see Figs. 34a, 34b & 47b.)

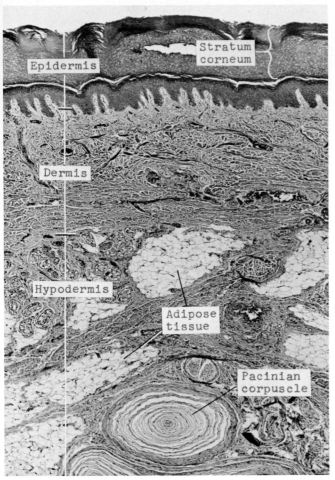

Fig. 48a   Pacinian corpuscle in hypodermis beneath thick skin x.s. x40. Pacinian corpuscles are pressure receptors. (Also see Figs. 33c & 33d.)

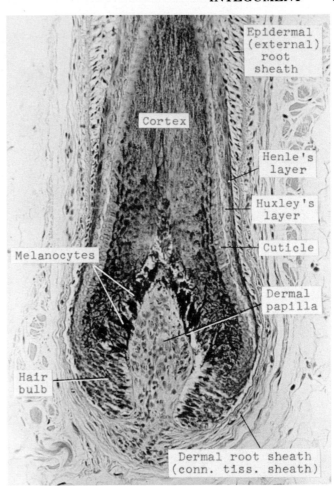

Fig. 48b   Hair bulb l.s. x430.

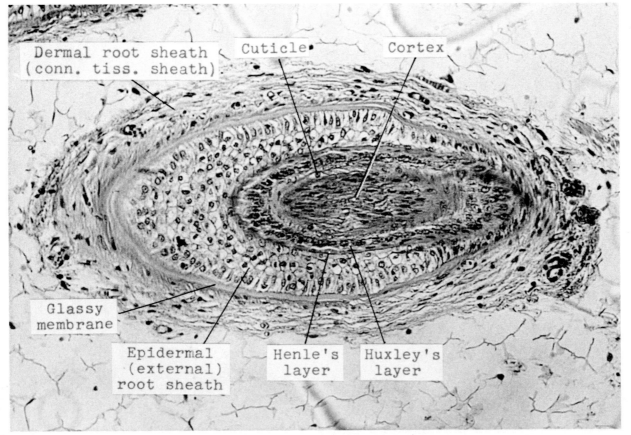

Fig. 48c   Hair follicle, tangential section x430.

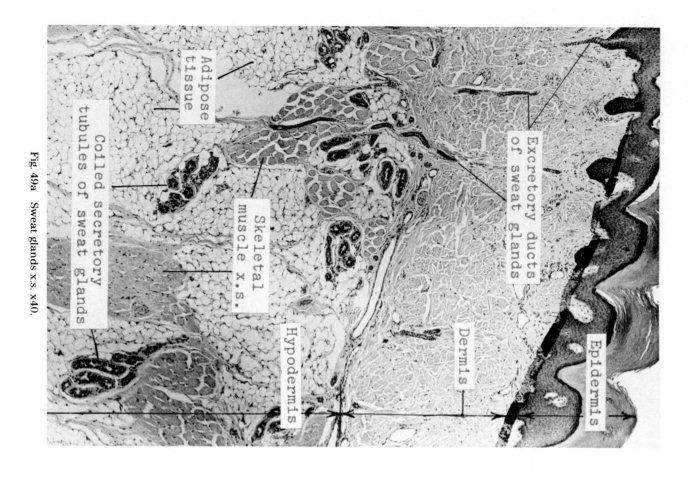

Fig. 49a   Sweat glands x.s. x40.

Adipose tissue

Coiled secretory tubules of sweat glands

Skeletal muscle x.s.

Hypodermis

Excretory ducts of sweat glands

Dermis

Epidermis

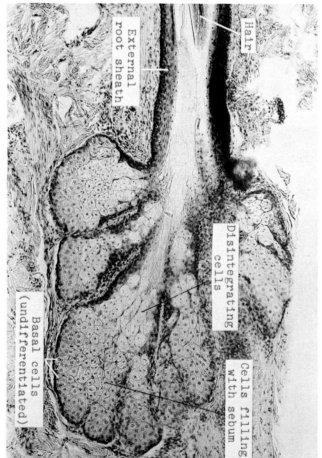

Fig. 49c   Sebaceous gland l.s. x100.

Hair

External root sheath

Disintegrating cells

Basal cells (undifferentiated)

Cells filling with sebum

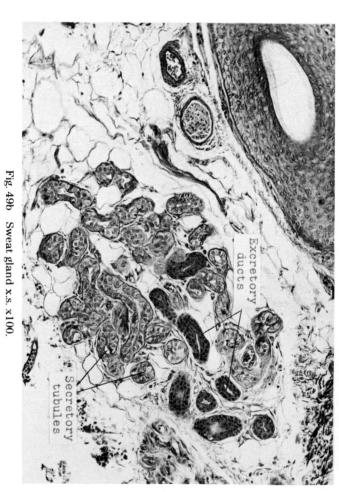

Fig. 49b   Sweat gland x.s. x100.

Excretory ducts

Secretory tubules

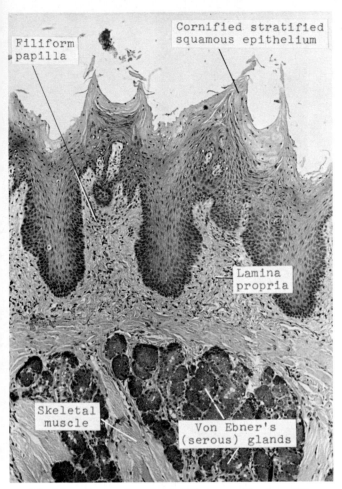

Fig. 50a   Tongue (monkey), filiform papillae l.s. x100.

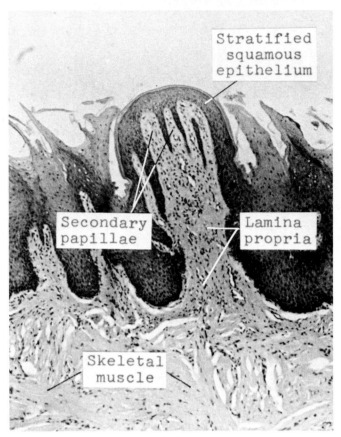

Fig. 50b   Tongue (monkey), fungiform papilla l.s. x40. Fungiform papillae often contain taste buds on the surface.

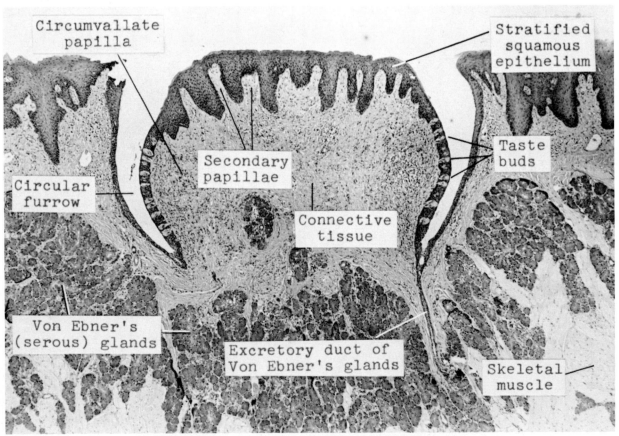

Fig. 50c   Tongue (monkey), circumvallate papilla l.s. x40.

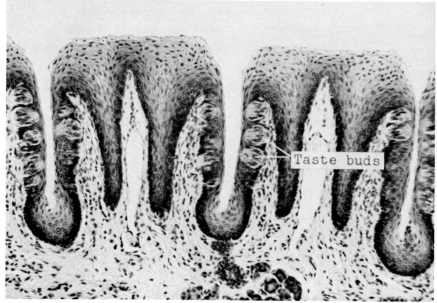

Fig. 51a   Tongue (rabbit), foliate papillae l.s. x100.

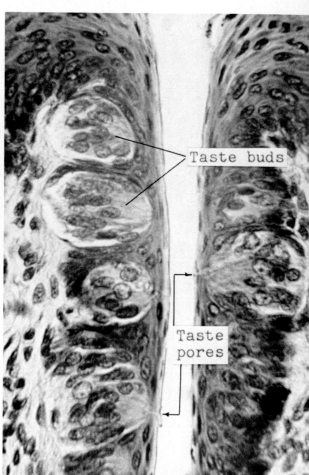

Fig. 51b   Tongue (rabbit), taste buds in foliate papillae l.s. x430. Notice the taste pores (entrances) into the taste buds.

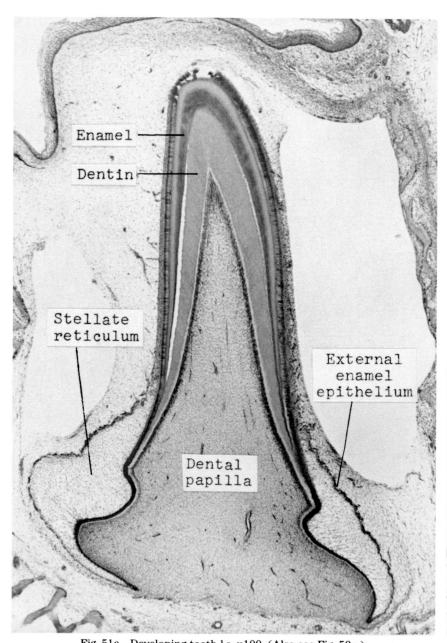

Fig. 51c   Developing tooth l.s. x100. (Also see Fig. 53a.)

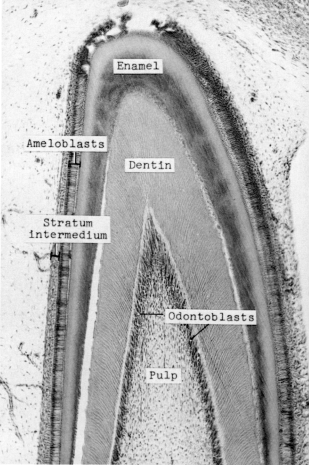

Fig. 51d   Developing tooth l.s. x100. (Also see Fig. 53a.)

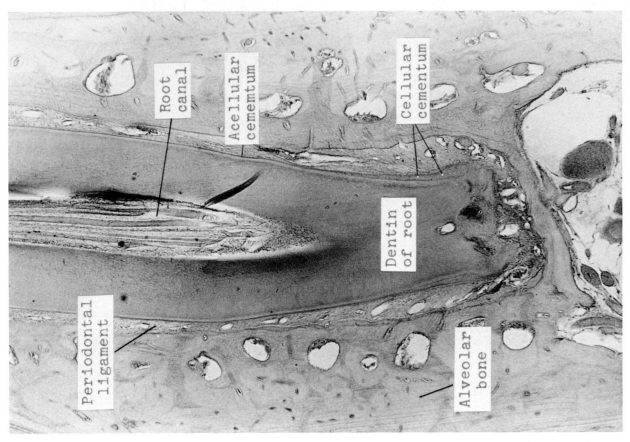

Fig. 52b Tooth, incisor l.s. x40.

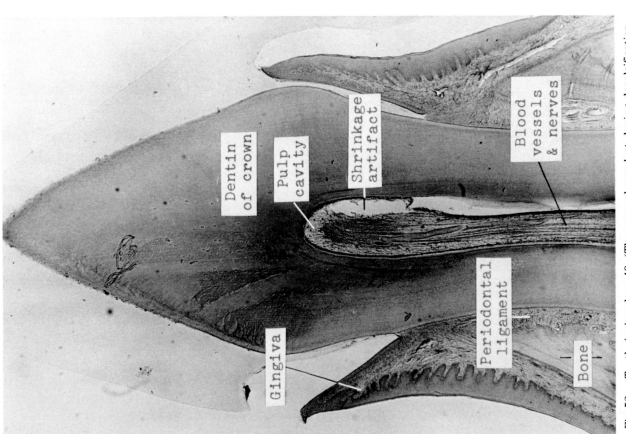

Fig. 52a Tooth, incisor l.s. x40. (The enamel was lost during decalcification prior to the staining process.)

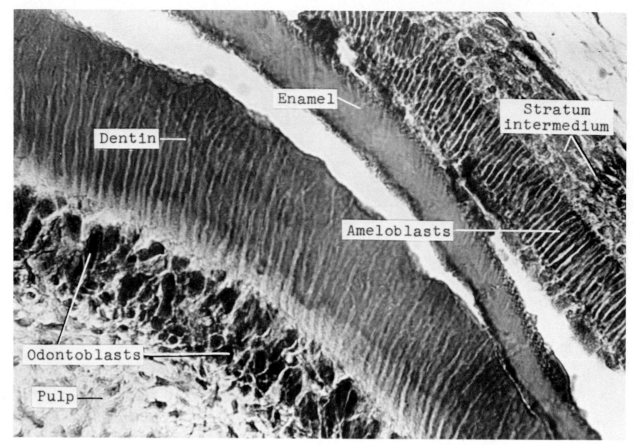

Fig. 53a   Developing tooth, crown l.s. x100. (Also see Figs. 51c & 51d.)

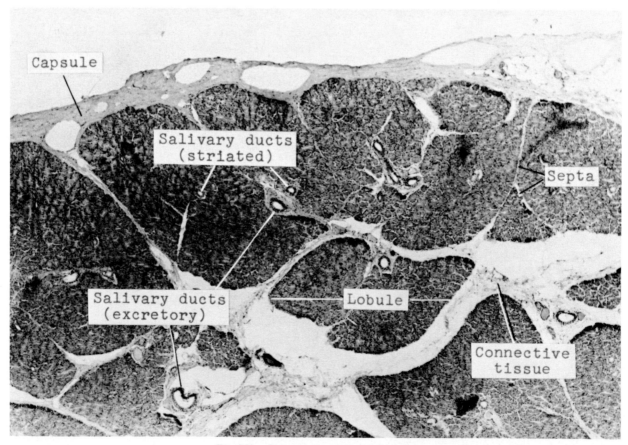

Fig. 53b   Parotid gland (cat) x.s. x40.

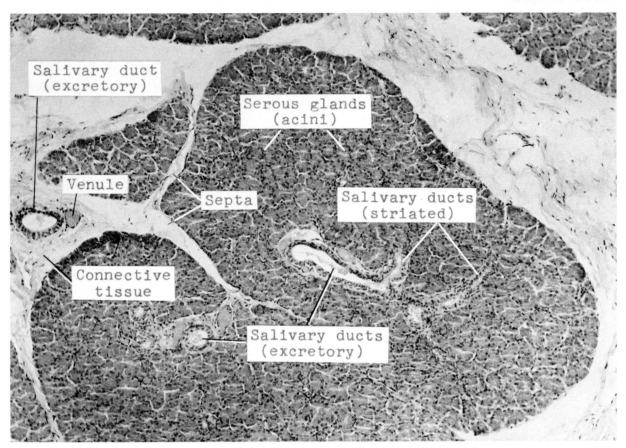

Fig. 54a   Parotid gland (cat) x.s. x100. Notice that the acini are all serous (protein producing). The acini produce the enzyme salivary amylase and other proteins.

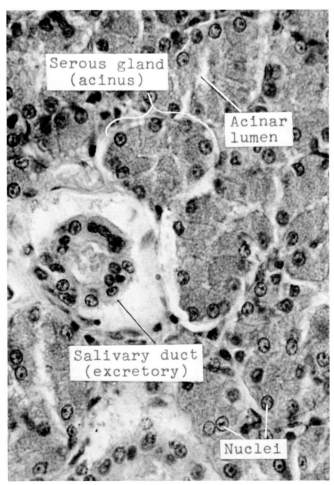

Fig. 54b   Parotid gland (cat) x.s. x430.

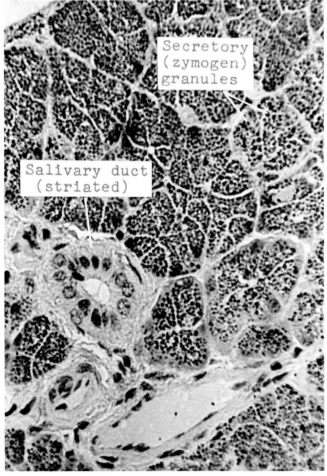

Fig. 54c   Parotid gland x.s. x430. (Special stain used to show the secretory granules which produce the digestive enzyme salivary amylase.)

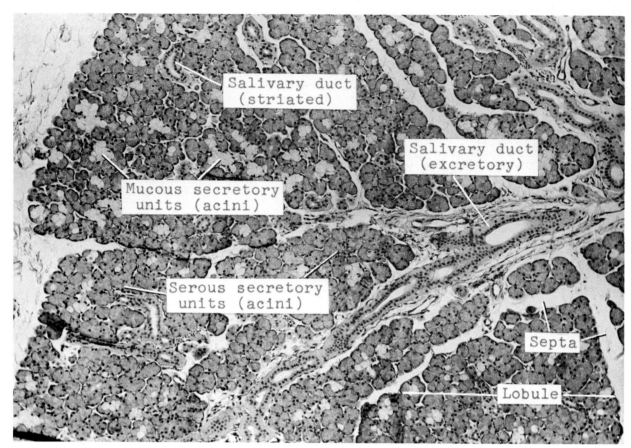

Fig. 55a   Submandibular gland (human) x.s. x100. The submandibular gland is a mixed gland containing both serous (protein producing) and mucous (mucus producing) acini.

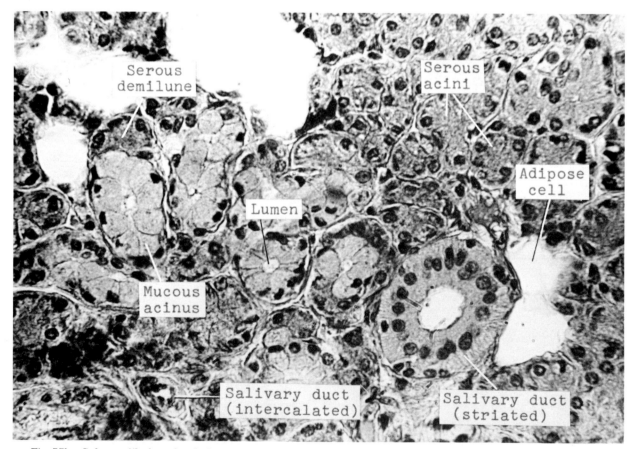

Fig. 55b   Submandibular gland (human) x.s. x430. Notice the mucous acinus with the serous demilune attached. (Also see Fig. 56c.)

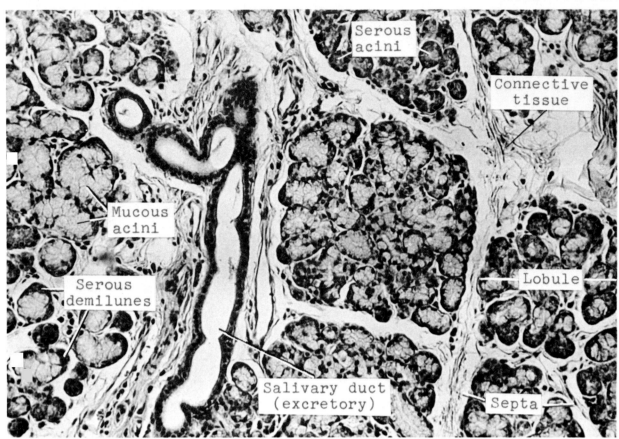

Fig. 56a   Sublingual gland x.s. x100. The sublingual gland consists mostly of mucous acini. Compare with the submandibular (Fig. 55a) and the parotid (Fig. 54a.)

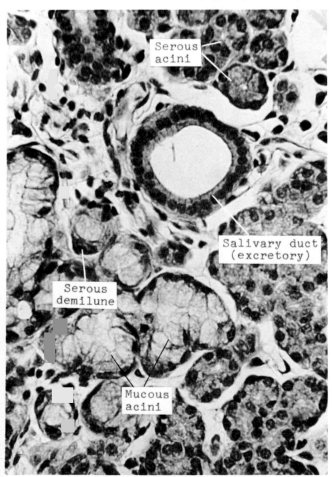

Fig. 56b   Sublingual gland x.s. x430.

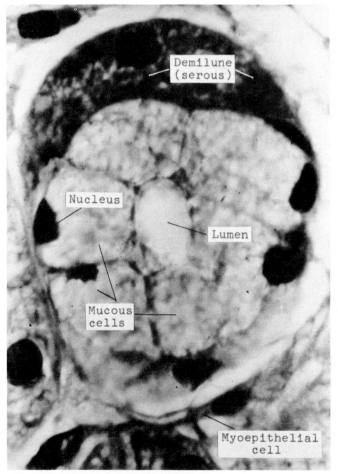

Fig. 56c   Mixed secretory unit in the parotid gland x.s. x1000. Note the serous demilune attached to the mucous acinus. (Also see Fig. 55b.) The myoepithelial (basket) cells are contractile and help move the secretions into the ducts.

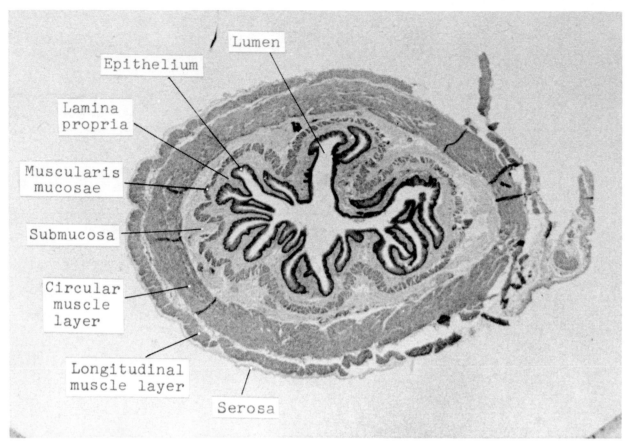

Fig. 57a   Esophagus (cat) x.s. x20.

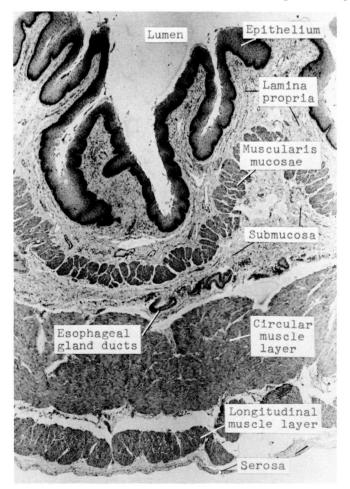

Fig. 57b   Esophagus (cat) x.s. x40.

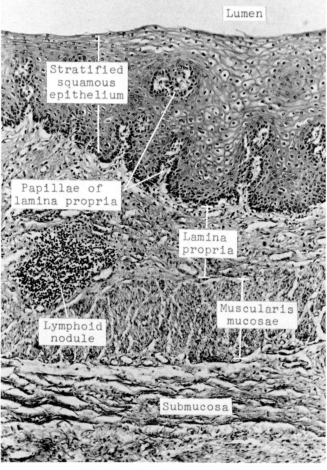

Fig. 57c   Esophagus x.s. x100.

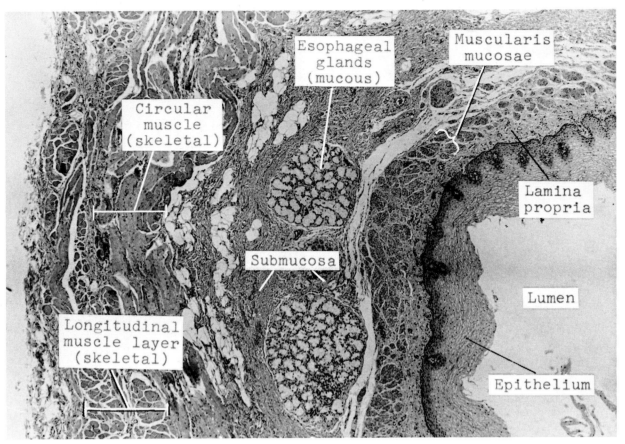

Fig. 58a   Upper esophagus x.s. x40.

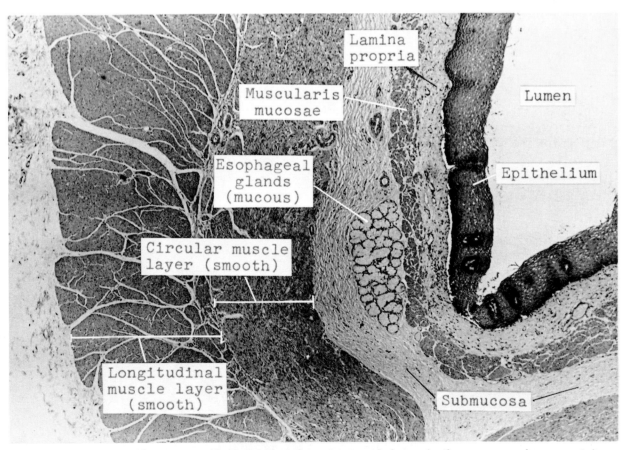

Fig. 58b   Lower esophagus x.s. x40. Notice that the outer muscle layers in the upper esophagus contain
skeletal (voluntary) muscle, while in the lower esophagus these layers are smooth (involuntary).
This allows you to initiate swallowing voluntarily and then allow the lower esophagus to take over
and finish the job without further effort on your part by involuntary peristalsis.

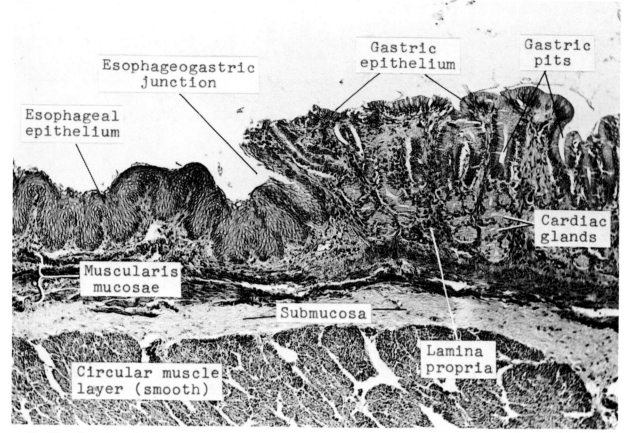

Fig. 59a   Junction of the esophagus and cardiac stomach l.s. x100. Notice the abrupt change from esophageal to gastric mucosa.

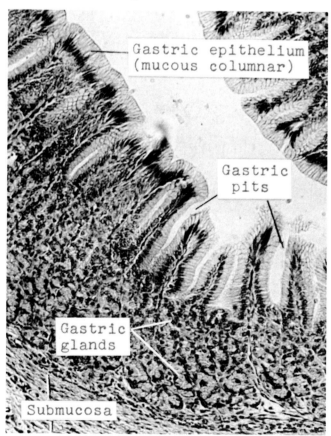

Fig. 59b   Stomach, cardiac region (cat) x.s. x100. Cardiac glands secrete primarily mucus.

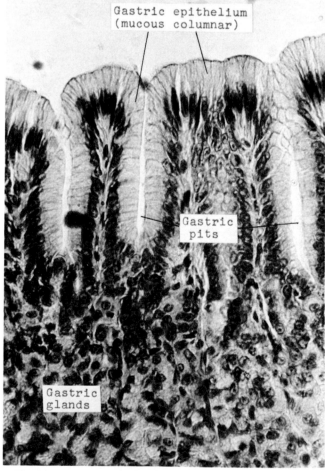

Fig. 59c   Stomach, cardiac region (cat) x.s. x430. Each epithelial cell in the gastric surface and pits secretes mucus.

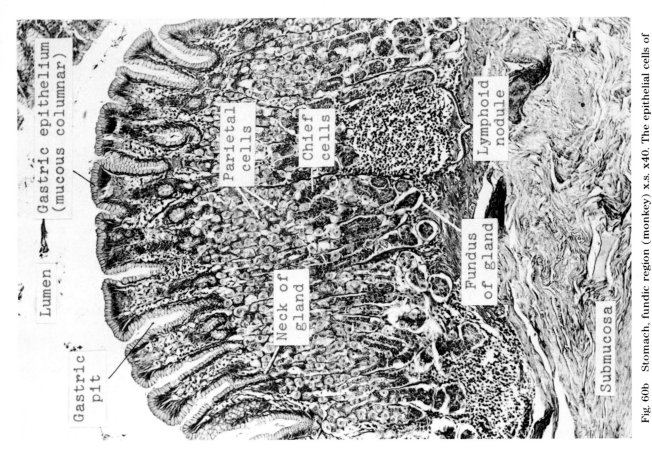

Fig. 60b Stomach, fundic region (monkey) x.s. x40. The epithelial cells of the surface and pits secrete mucus. The fundic glands secrete hydrochloric acid and the digestive enzyme pepsinogen. (See Fig. 61a.)

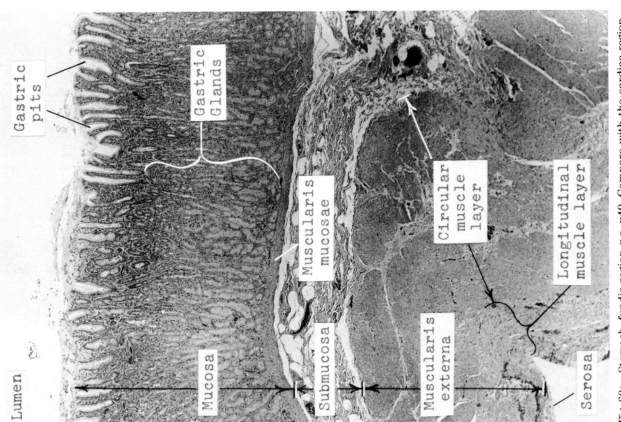

Fig. 60a Stomach, fundic region x.s. x40. Compare with the cardiac region (Fig. 59b) and the pyloric region (Fig. 61d). Notice how much shallower the gastric pits are in the fundic region (extending into about one fourth of the mucosa).

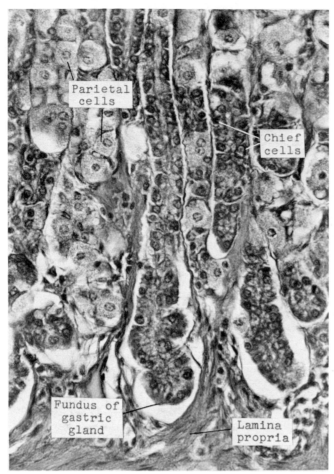

Fig. 61a   Fundic glands (monkey) l.s. x430. Parietal cells secrete hydrochloric acid and chief cells synthesize pepsinogen.

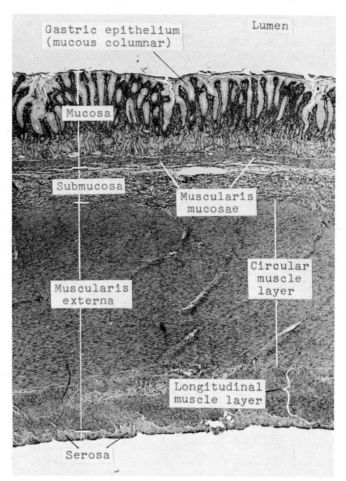

Fig. 61b   Stomach, pyloric region (monkey) x.s. x40.

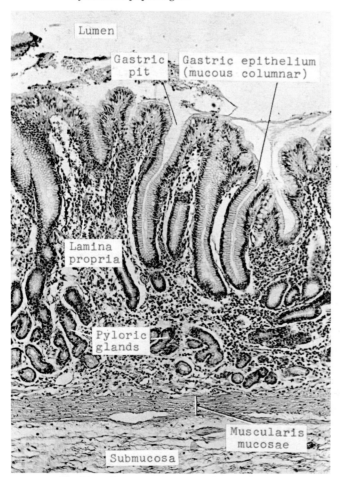

Fig. 61c   Stomach, pyloric region x.s. x100.

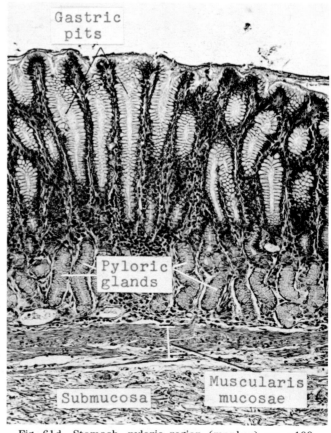

Fig. 61d   Stomach, pyloric region (monkey) x.s. x100. Pyloric glands produce mucus and gastrin.

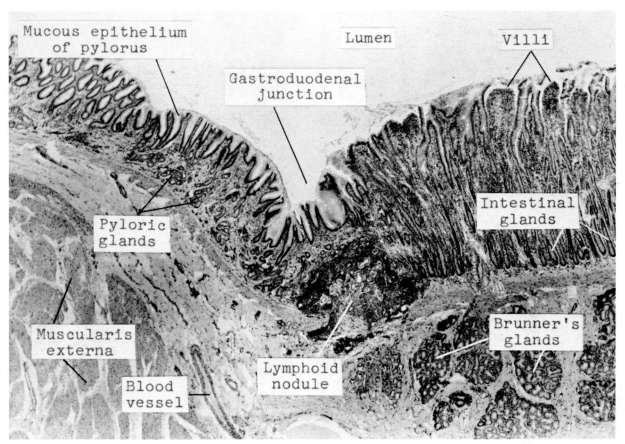

Fig. 62a   Junction of the stomach and duodenum l.s. x40. Notice the abrupt change from gastric to duodenal mucosa and the sudden appearance of Brunner's glands in the submucosa.

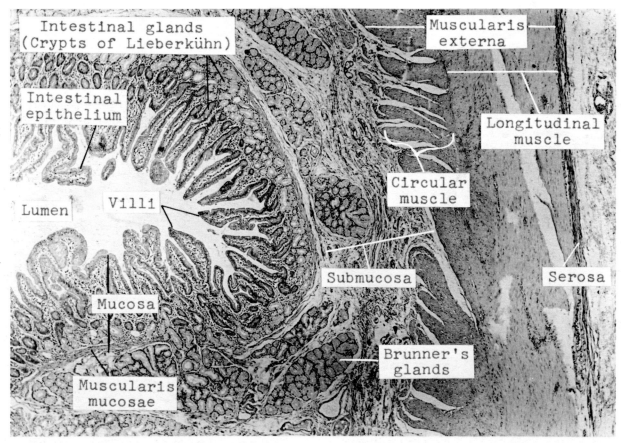

Fig. 62b   Duodenum x.s. x40. Brunner's glands are found only in the duodenum and produce mucus (a protection against ulceration) and alkaline secretions (to neutralize the acidic bolus from the stomach).

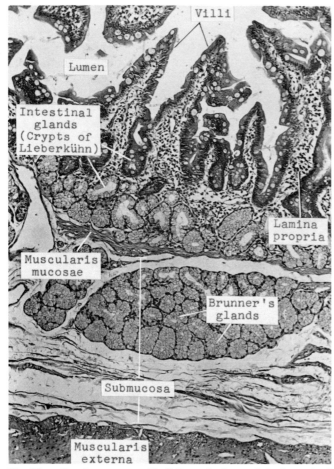

Fig. 63a   Duodenum x.s. x40. Cells at the base of the intestinal glands divide frequently to form cells that migrate up the villi to replace exfoliated surface cells.

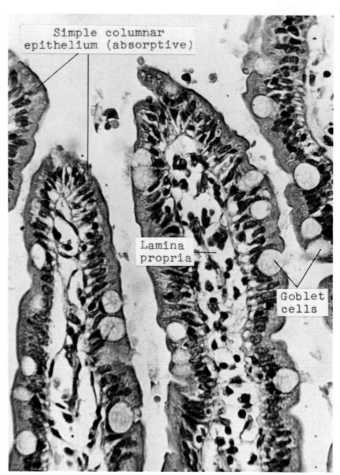

Fig. 63b   Villi (duodenum) l.s. x430. The epithelium consists mostly of goblet cells and absorptive cells.

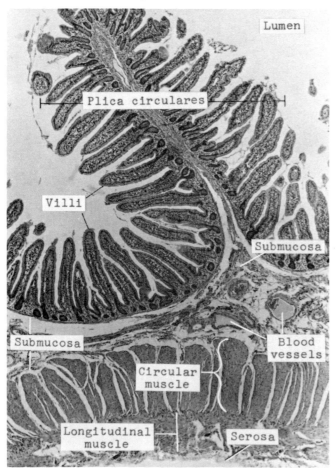

Fig. 63c   Jejunum x.s. x40. The plica circulares is a large fold of mucosa.

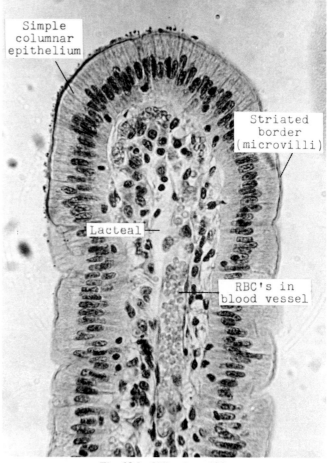

Fig. 63d   Villus l.s. x430.

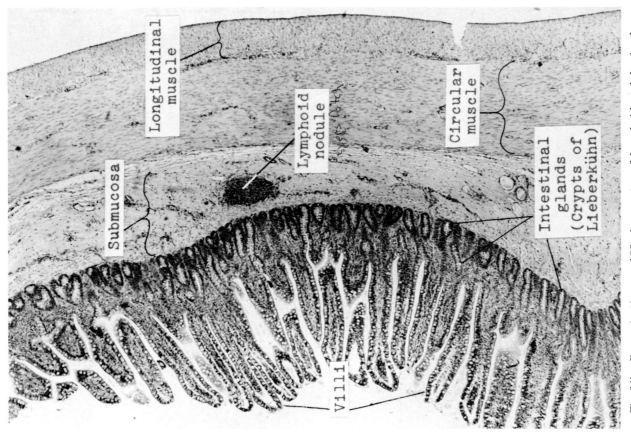

Fig. 64b   Ileum (cat) x.s. x100. Aggregates of lymphoid nodules in the ileum are called Peyer's patches and are a characteristic feature of the ileum.

Longitudinal muscle

Lymphoid nodule

Circular muscle

Submucosa

Intestinal glands (Crypts of Lieberkühn)

Villi

Peyer's Patch

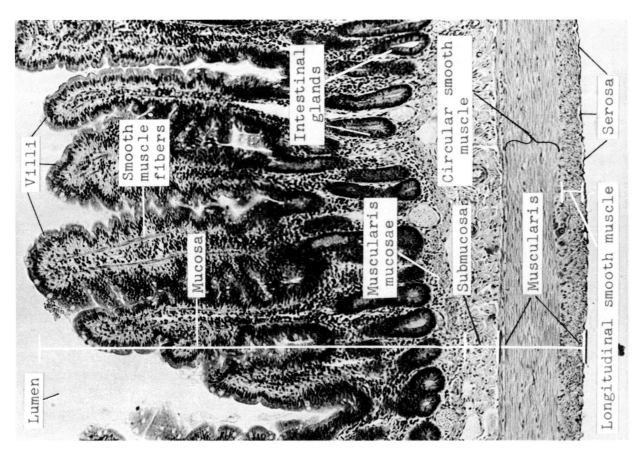

Fig. 64a   Small intestine, region of the jejunum-ileum x.s. x100.

Villi

Smooth muscle fibers

Intestinal glands

Circular smooth muscle

Mucosa

Muscularis mucosae

Submucosa

Muscularis

Serosa

Longitudinal smooth muscle

Lumen

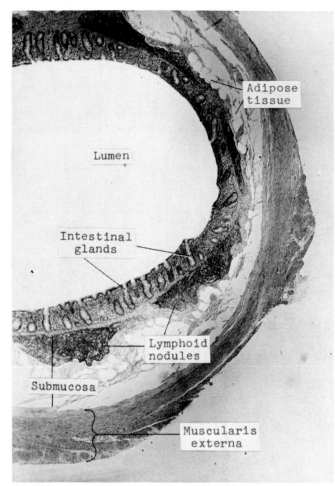

Fig. 65a   Veriform appendix x.s. x40.

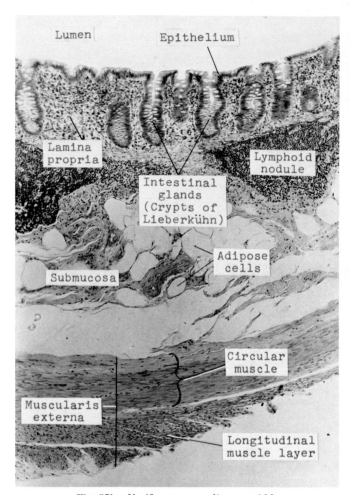

Fig. 65b   Veriform appendix x.s. x100.

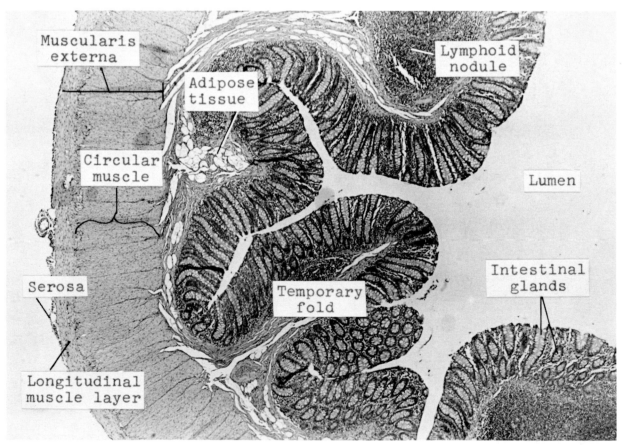

Fig. 65c   Colon x.s. x40.

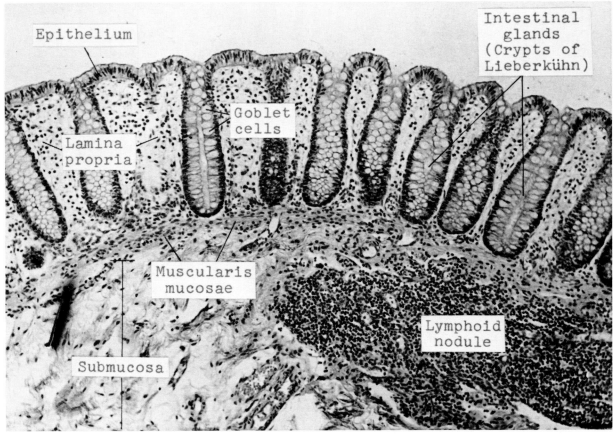

Fig. 66a   Colon x.s. x100. The colon has no villi, but it does have these crypts.

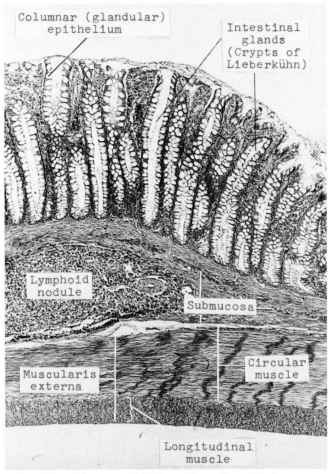

Fig. 66b   Colon x.s. x100. The goblet cells tend to increase in number closer to the rectum.

Fig. 66c   Intestinal glands (crypts) x.s. x430. This view of the crypts is what you see if a section were cut horizontally across the crypts in Fig. 66b. The absorptive cells absorb fluids and the goblet cells produce mucus for lubrication of the intestinal surface.

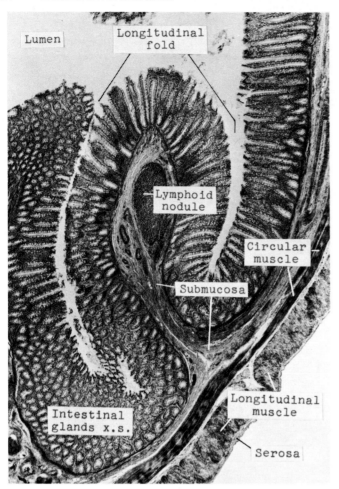

Lumen

Longitudinal fold

Lymphoid nodule

Circular muscle

Submucosa

Intestinal glands x.s.

Longitudinal muscle

Serosa

Fig. 67a   Rectum x.s. x40.

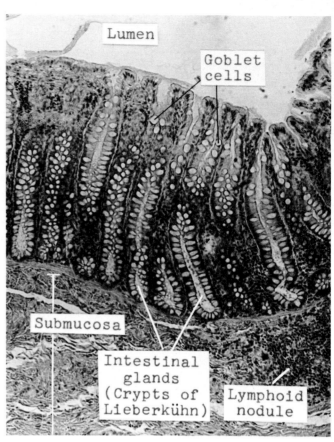

Lumen

Goblet cells

Submucosa

Intestinal glands (Crypts of Lieberkühn)

Lymphoid nodule

Fig. 67b   Rectum x.s. x100.

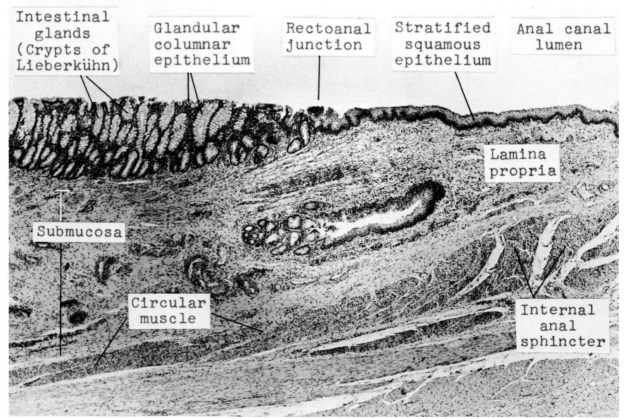

Intestinal glands (Crypts of Lieberkühn)

Glandular columnar epithelium

Rectoanal junction

Stratified squamous epithelium

Anal canal lumen

Lamina propria

Submucosa

Circular muscle

Internal anal sphincter

Fig. 67c   Junction of the rectum and anal canal l.s. x40.

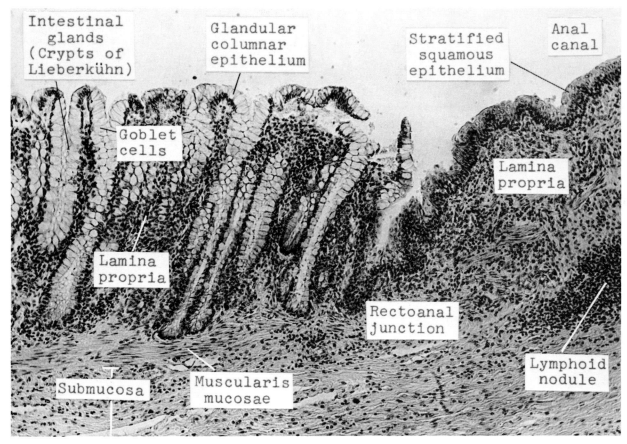

Fig. 68a Junction of the rectum and anal canal l.s. x100.

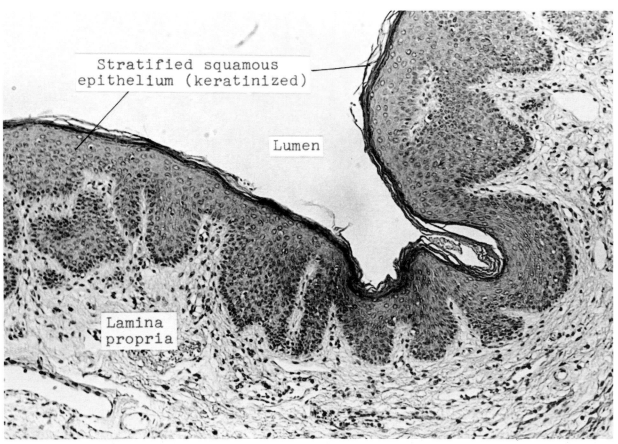

Fig. 68b Anal canal x.s. x430.

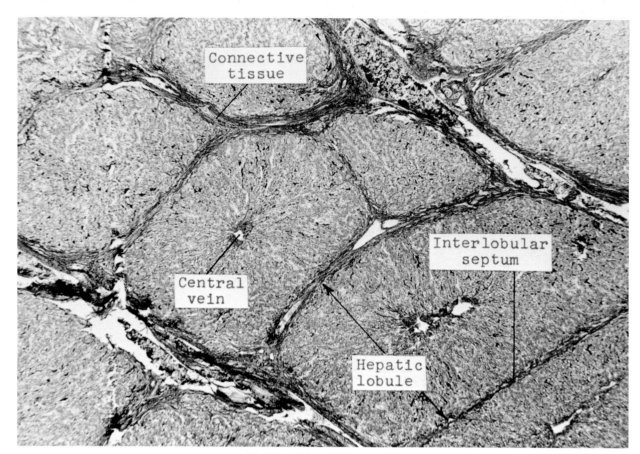

Fig. 69a   Liver (pig) x.s. x40.

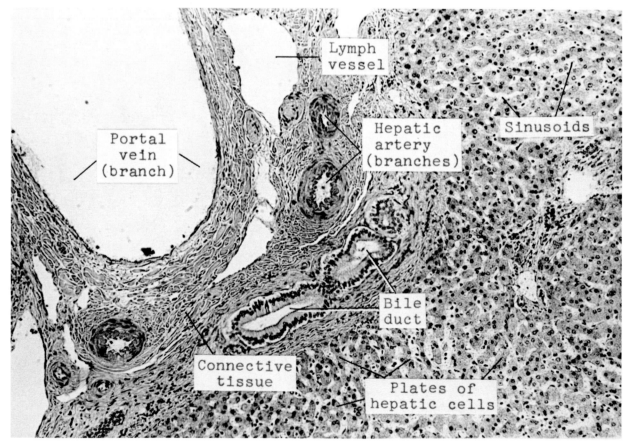

Fig. 69b   Liver x.s. x100.

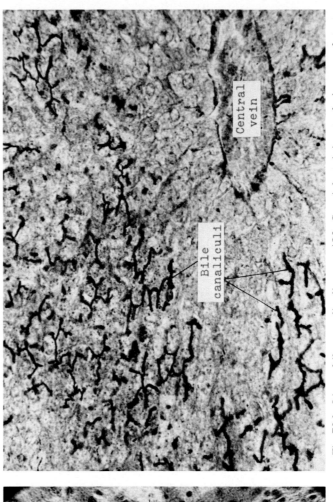

Fig. 70b Liver, center of lobule x.s. x430.

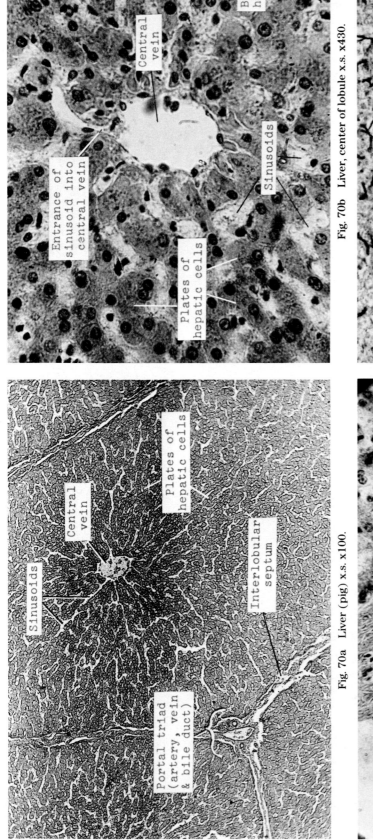

Fig. 70a Liver (pig) x.s. x100.

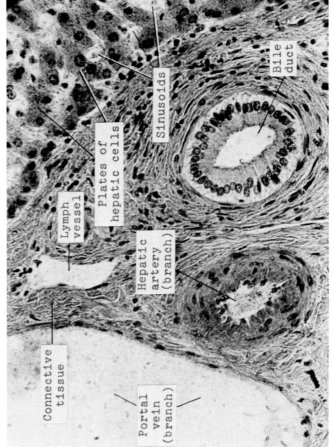

Fig. 70d Liver, bile canaliculi x.s. x430. Special staining technique used to illustrate the canaliculi.

Fig. 70c Liver, portal triad (artery, vein & bile duct) x.s. x430.

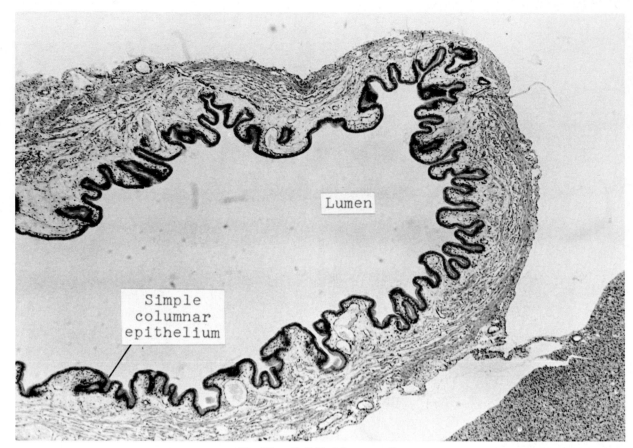

Fig. 71a   Gall bladder x.s. x40.

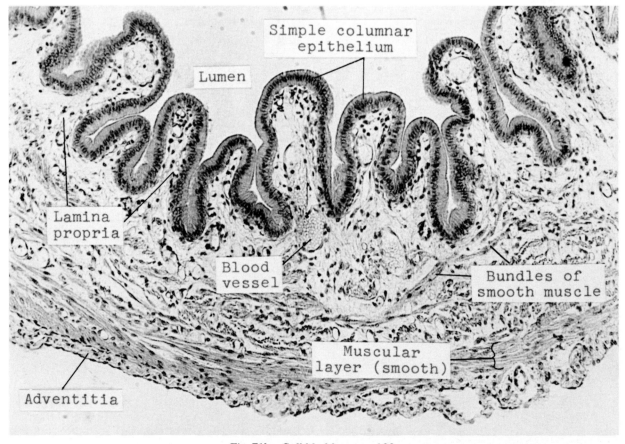

Fig. 71b   Gall bladder x.s. x100.

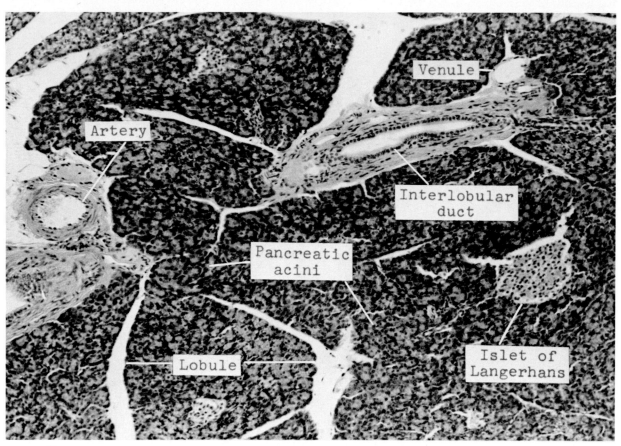

Fig. 72a   Pancreas x.s. x100. The pancreatic acini make up the bulk of the exocrine pancreas. The Islets of Langerhans form the endocrine pancreas.

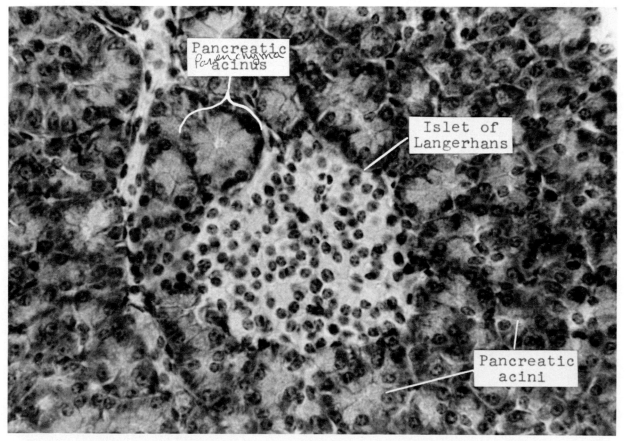

Fig. 72b   Pancreas, islet of Langerhans x.s. x430. The acini produce digestive enzymes while the islets of Langerhans produce several hormones including insulin.

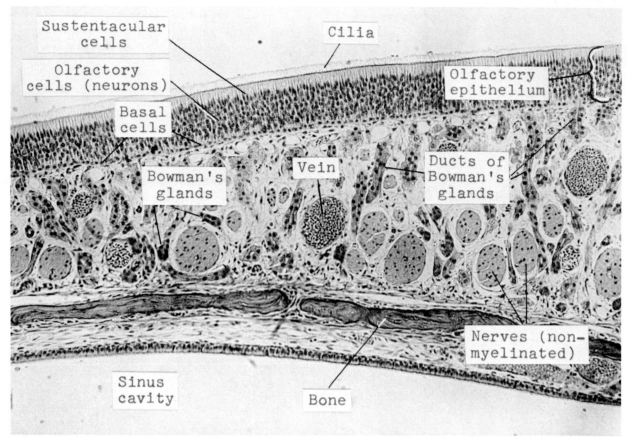

Fig. 73a   Olfactory mucosa x.s. x100.

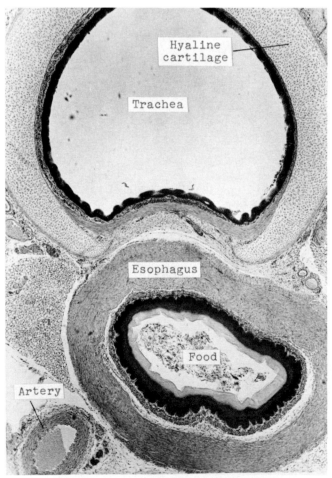

Fig. 73b   Trachea and esophagus x.s. x40.

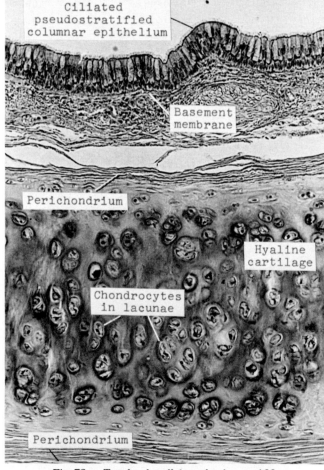

Fig. 73c   Tracheal wall (monkey) x.s. x100.

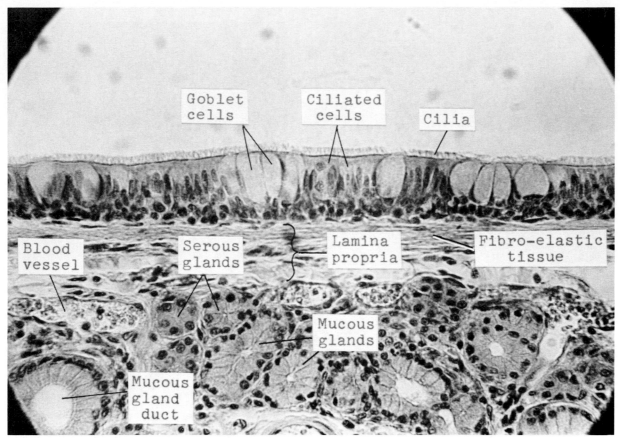

Goblet cells

Ciliated cells

Cilia

Blood vessel

Serous glands

Lamina propria

Fibro-elastic tissue

Mucous glands

Mucous gland duct

Fig. 74a   Trachea, ciliated pseudostratified columnar epithelial cells l.s. x430. These cells are found lining the respiratory passages (mucous membranes of nose, trachea and bronchii). The goblet cells produce mucus which catches dirt particles and the cilia beat and drive the dirt-laden mucus to the back of the throat where it is swallowed. Smoking paralyzes these cilia. The smoker must then cough to remove the mucus in his lung passages.

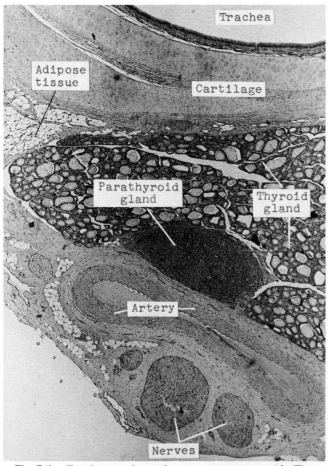

Trachea

Adipose tissue

Cartilage

Parathyroid gland

Thyroid gland

Artery

Nerves

Fig. 74b   Trachea and nearby structures x.s. x40. The artery is probably the carotid and the nerve is most likely the vagus.

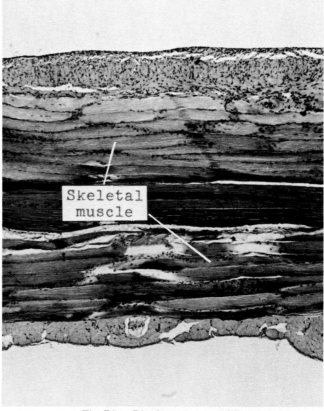

Skeletal muscle

Fig. 74c   Diaphragm x.s. x100.

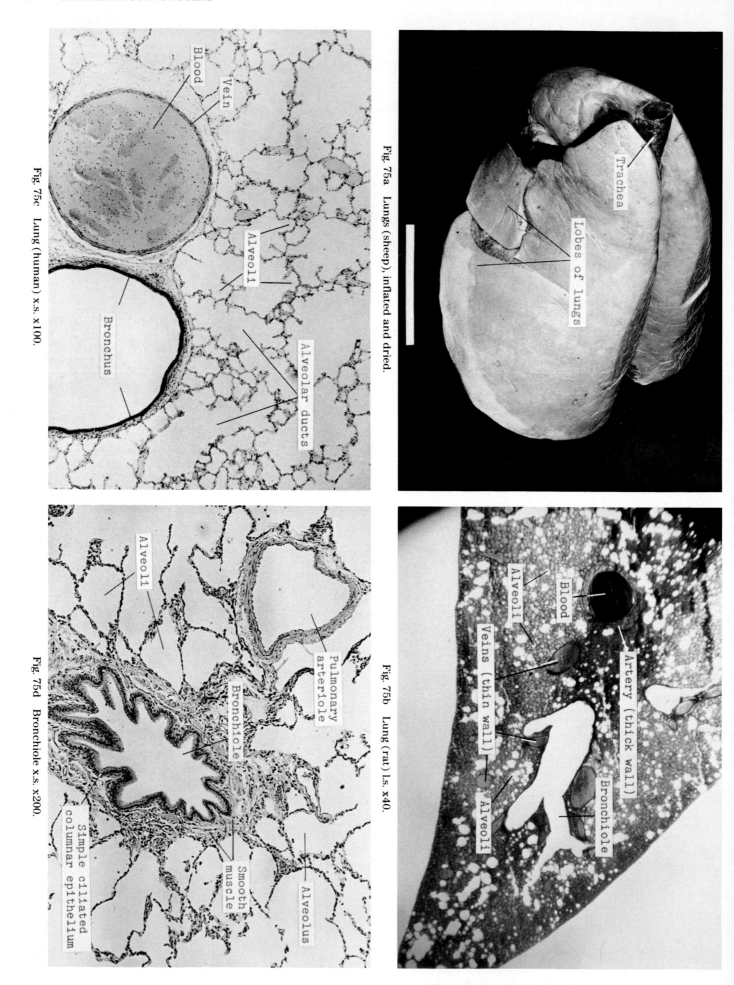

Fig. 75c   Lung (human) x.s. x100.

Blood

Vein

Alveoli

Bronchus

Alveolar ducts

Fig. 75a   Lungs (sheep), inflated and dried.

Trachea

Lobes of lungs

Fig. 75d   Bronchiole x.s. x200.

Alveoli

Pulmonary arteriole

Bronchiole

Simple ciliated columnar epithelium

Smooth muscle

Alveolus

Fig. 75b   Lung (rat) l.s. x40.

Alveoli

Blood

Veins (thin wall)

Artery (thick wall)

Alveoli

Bronchiole

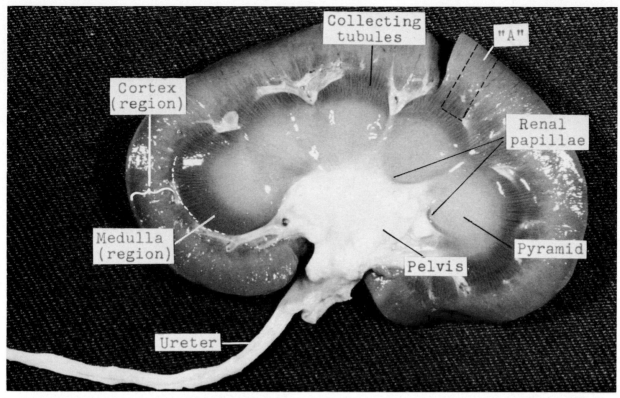

Fig. 76a   Fresh kidney (lamb) dissection l.s. x1.

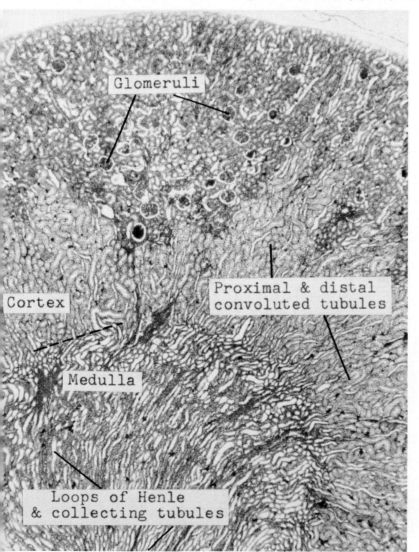

Fig. 76b   Kidney (rat) l.s. x40. (From an area similar to region "A" in Fig. 76a.)

Fig. 76c   Kidney l.s. x20.

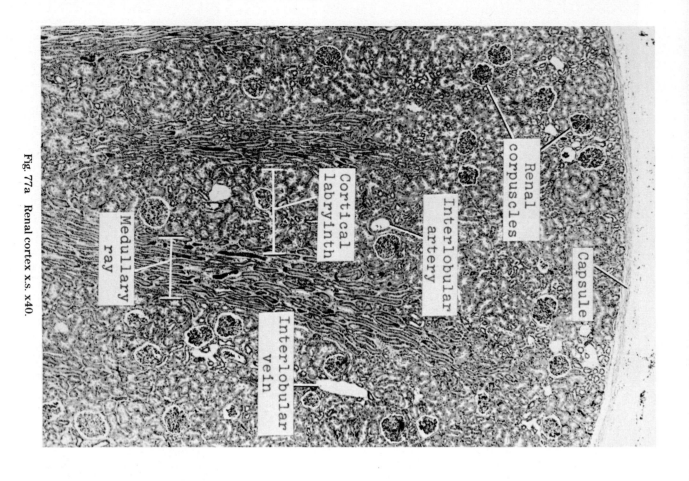

Fig. 77a   Renal cortex x.s. x40.

Renal corpuscles

Cortical labryinth

Medullary ray

Interlobular artery

Interlobular vein

Capsule

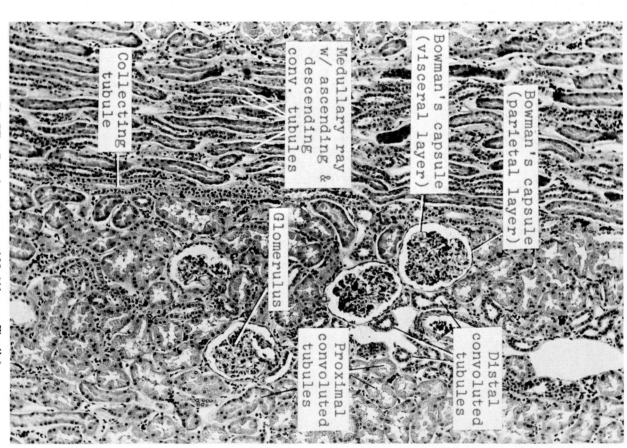

Fig. 77b   Renal cortex x.s. x100. (Also see Fig. 4b.)

Collecting tubule

Medullary ray w/ ascending & descending conv. tubules

Bowman's capsule (visceral layer)

Bowman's capsule (parietal layer)

Glomerulus

Proximal convoluted tubules

Distal convoluted tubules

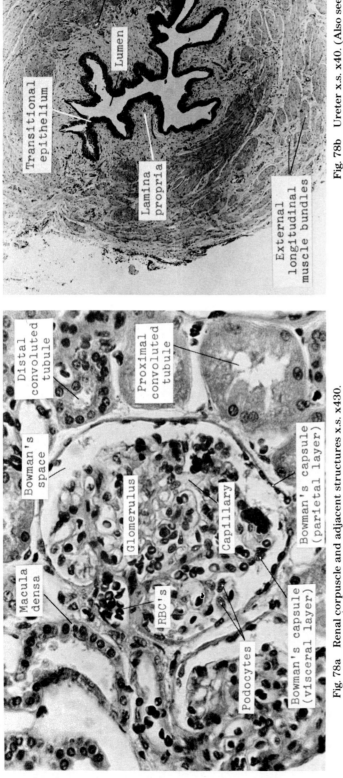

Fig. 78b   Ureter x.s. x40. (Also see Fig. 78c.)

Adventitial connective tissue

Internal longitudinal muscle bundles

Circular muscle bundles

Lumen

Transitional epithelium

Lamina propria

External longitudinal muscle bundles

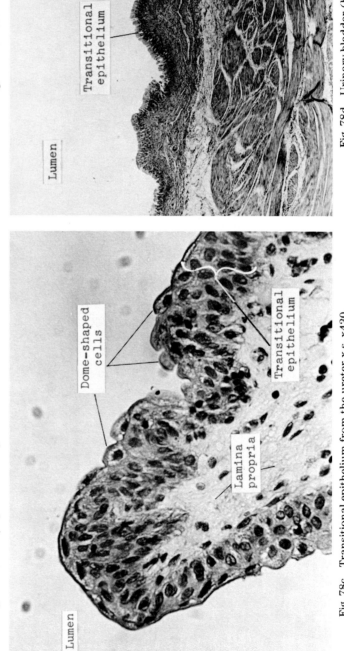

Fig. 78d   Urinary bladder (human) x.s. x40. (Also see Fig. 4a.)

Lamina propria

Longitudinal smooth muscle bundles

Circular smooth muscle bundles

Transitional epithelium

Lumen

Fig. 78a   Renal corpuscle and adjacent structures x.s. x430.

Distal convoluted tubule

Proximal convoluted tubule

Bowman's space

Macula densa

Glomerulus

Capillary

Bowman's capsule (parietal layer)

RBC's

Podocytes

Bowman's capsule (visceral layer)

Fig. 78c   Transitional epithelium from the ureter x.s. x430.

Dome-shaped cells

Transitional epithelium

Lamina propria

Lumen

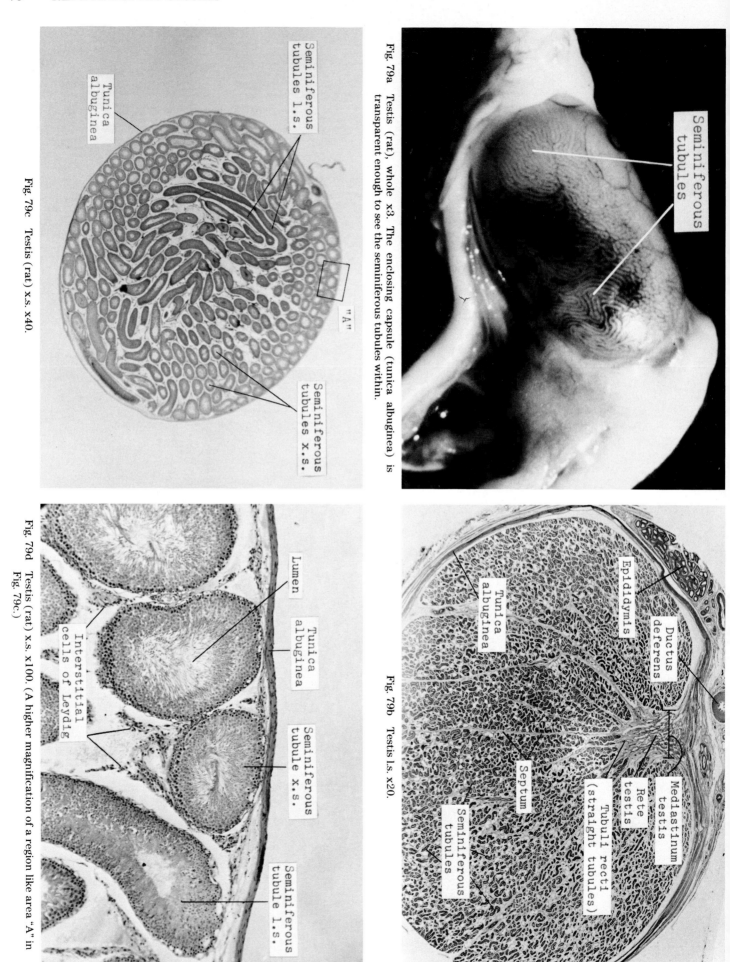

Fig. 79a Testis (rat), whole x3. The enclosing capsule (tunica albuginea) is transparent enough to see the seminiferous tubules within.

Seminiferous tubules

Fig. 79c Testis (rat) x.s. x40.

Tunica albuginea

Seminiferous tubules l.s.

"A"

Seminiferous tubules x.s.

Fig. 79b Testis l.s. x20.

Tunica albuginea

Epididymis

Ductus deferens

Septum

Mediastinum testis

Rete testis

Tubuli recti (straight tubules)

Seminiferous tubules

Fig. 79d Testis (rat) x.s. x100. (A higher magnification of a region like area "A" in Fig. 79c.)

Lumen

Tunica albuginea

Interstitial cells of Leydig

Seminiferous tubule x.s.

Seminiferous tubule l.s.

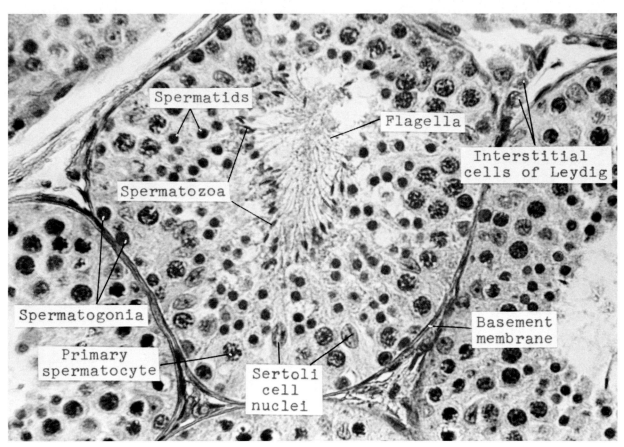

Fig. 80a   Seminiferous tubules x.s. x430.

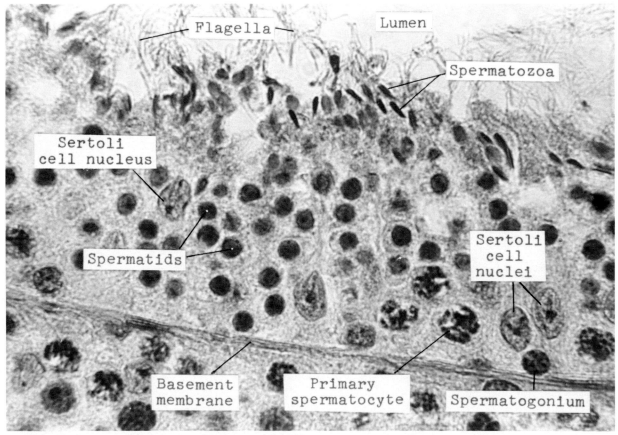

Fig. 80b   Seminiferous tubule x.s. x1000.

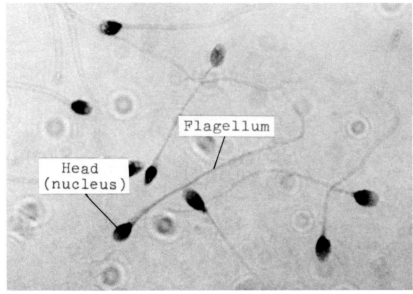

Fig. 81a Sperm (human) w.m. x1000.

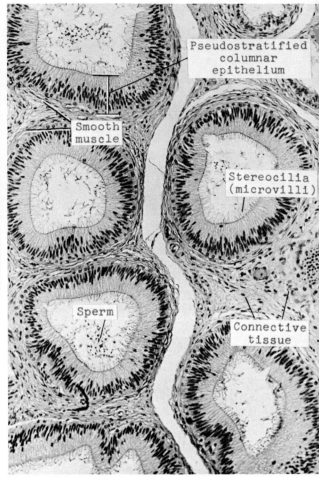

Fig. 81b Ductus epididymis x.s. x100.

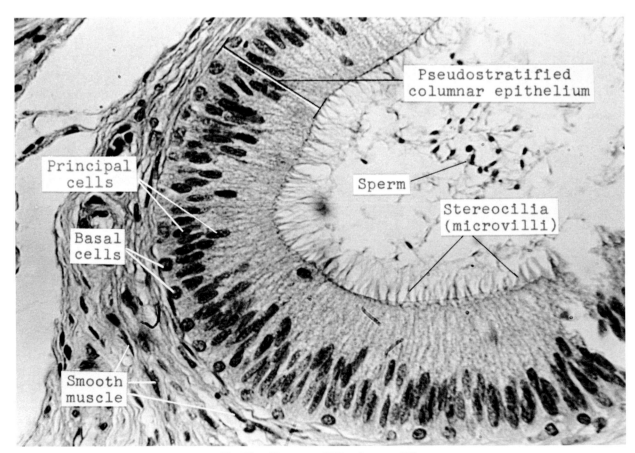

Fig. 81c Ductus epididymis x.s. x430.

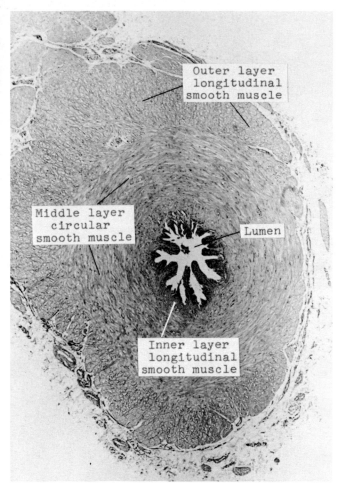

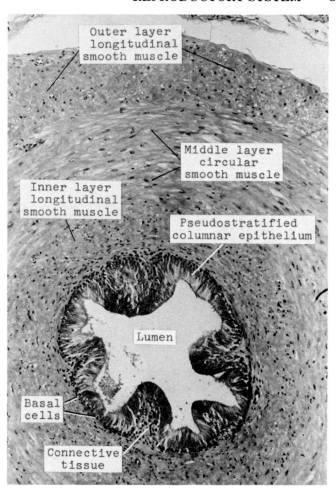

Fig. 82a   Ductus deferens (vas deferens) x.s. x40. The ductus deferens carries sperm from the epididymis to the ejaculatory duct.

Fig. 82b   Ductus deferens (vas deferens) x.s. x100.

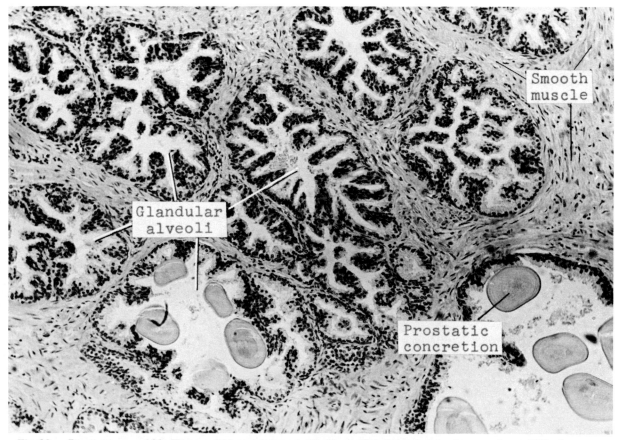

Fig. 82c   Prostate x.s. x100. The prostate secretes an alkaline milky fluid that makes up about 20% of semen. The prostatic concretions tend to increase in number with age and are not functional.

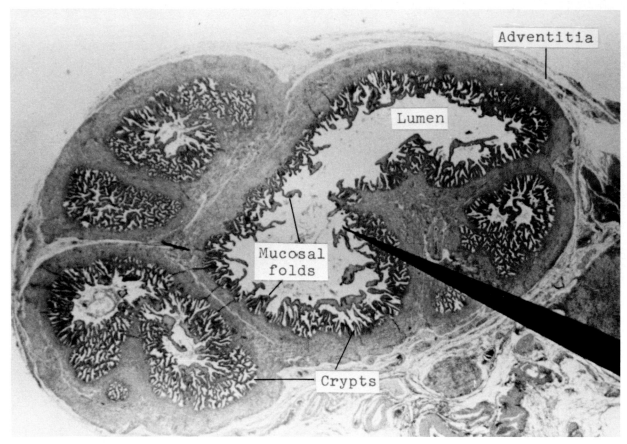

Fig. 83a   Seminal vesicle x.s. x40.

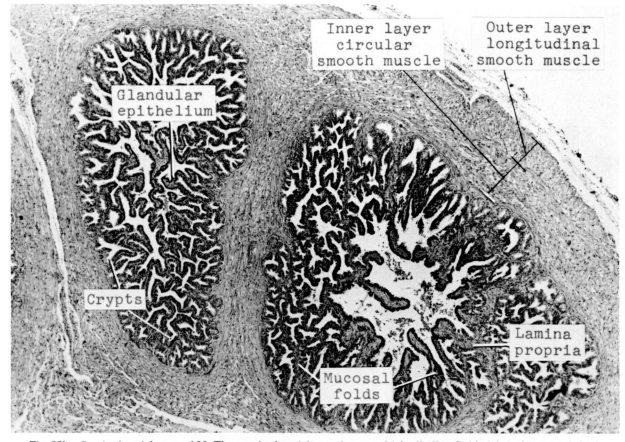

Fig. 83b   Seminal vesicle x.s. x100. The seminal vesicle produces a thick alkaline fluid rich in fructose (which nourishes the sperm) and makes up about 60% of the semen.

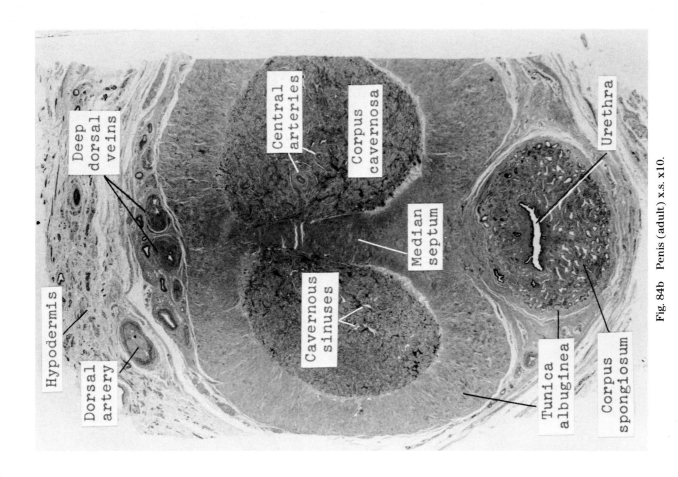

Deep dorsal veins

Central arteries

Corpus cavernosa

Urethra

Hypodermis

Dorsal artery

Cavernous sinuses

Median septum

Tunica albuginea

Corpus spongiosum

Fig. 84b Penis (adult) x.s. x10.

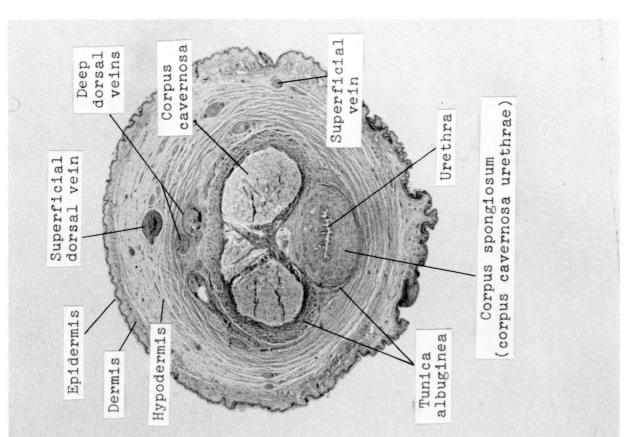

Deep dorsal veins

Corpus cavernosa

Superficial vein

Superficial dorsal vein

Urethra

Epidermis

Dermis

Hypodermis

Tunica albuginea

Corpus spongiosum (corpus cavernosa urethrae)

Fig. 84a Penis (infant) x.s. x10.

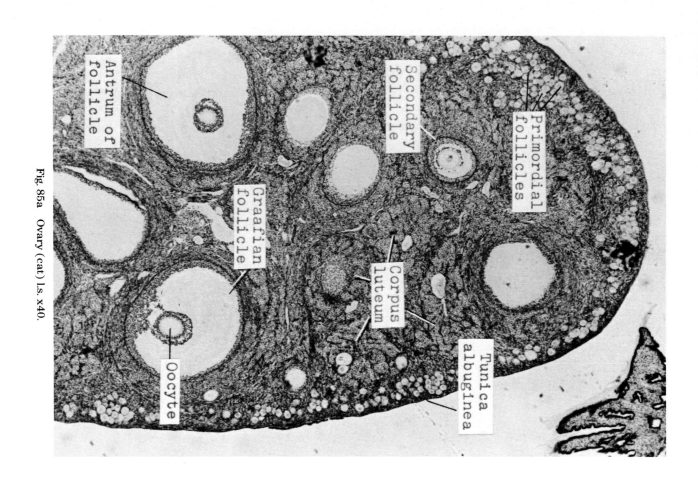

Fig. 85a   Ovary (cat) l.s. x40.

Antrum of follicle

Secondary follicle

Primordial follicles

Graafian follicle

Corpus luteum

Oocyte

Tunica albuginea

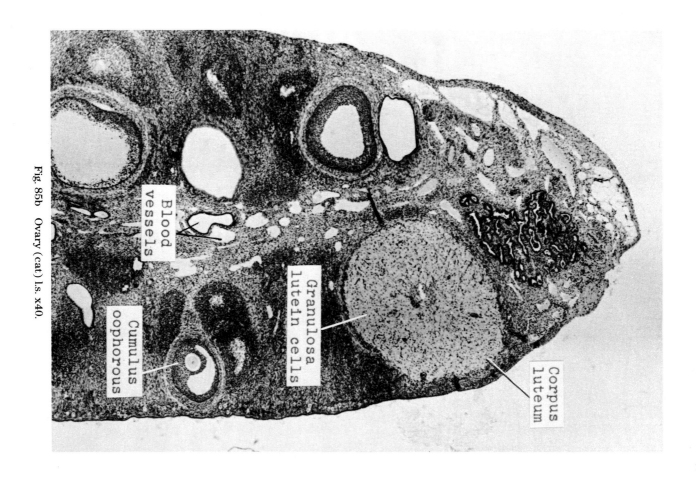

Fig. 85b   Ovary (cat) l.s. x40.

Blood vessels

Cumulus oophorous

Granulosa lutein cells

Corpus luteum

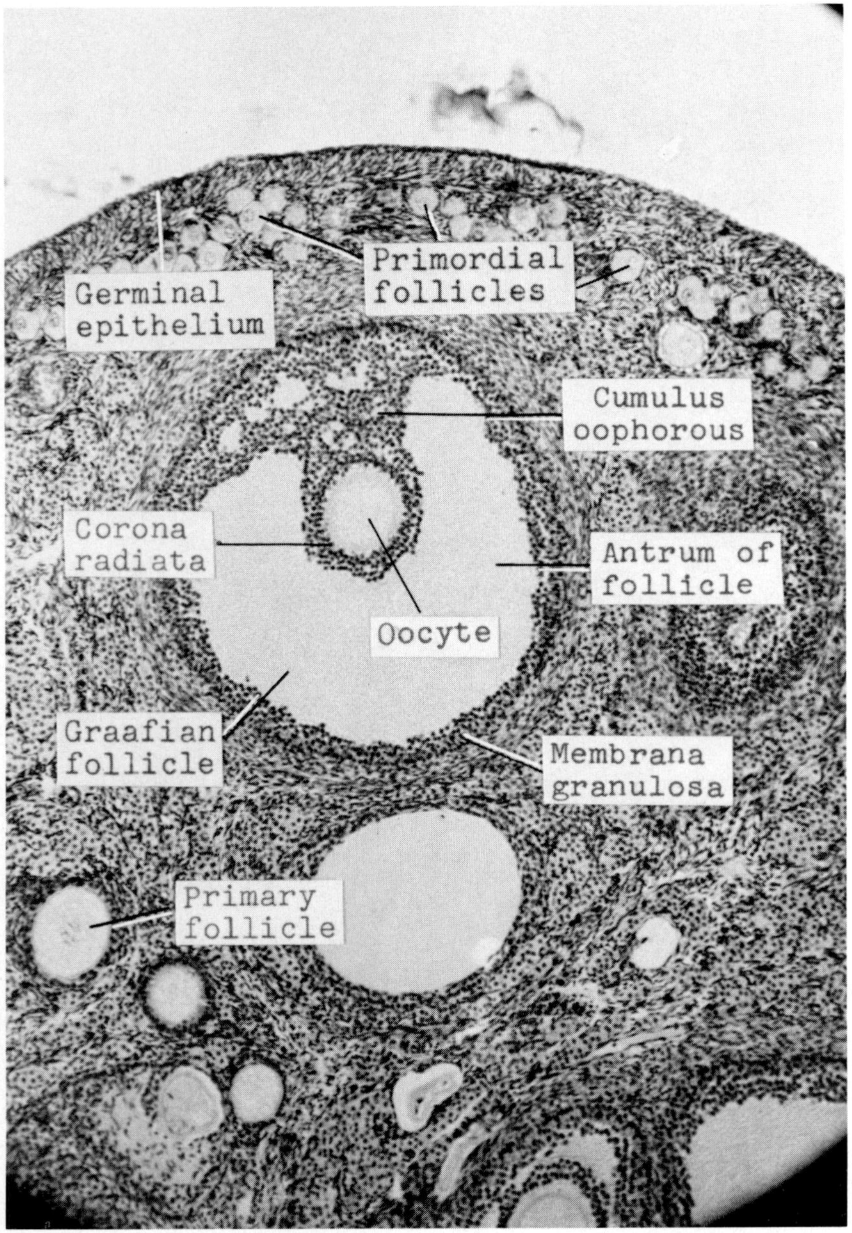

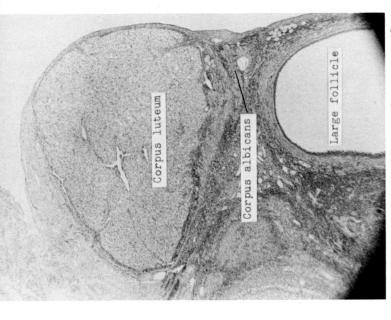

Fig. 86b   Ovary (cat) with corpus luteum x.s. x40.

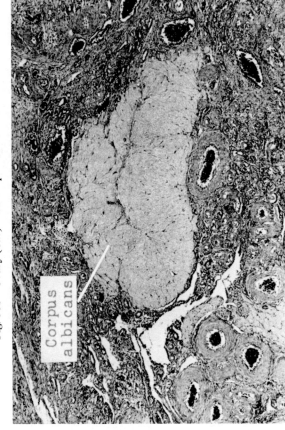

Fig. 86c   Corpus albicans x.s. x40. The corpus albicans is a mass of connective tissue and disintegrating cells from an old disintegrating corpus luteum which forms this non-functional white scar.

Fig. 86a   Ovary (cat) with Graafian follicle x.s. x60.

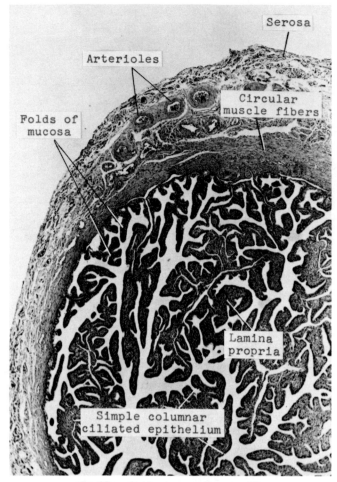

Fig. 87a   Oviduct, ampulla x.s. x40.

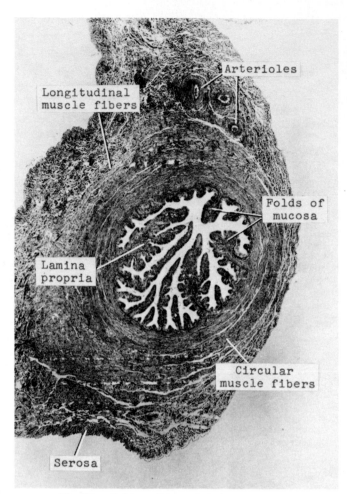

Fig. 87b   Oviduct (human), isthmus x.s. x40.

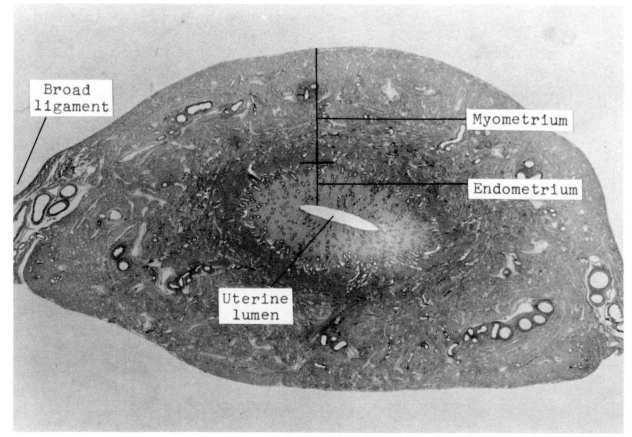

Fig. 87c   Uterus (monkey) x.s. x10. Proliferative phase.

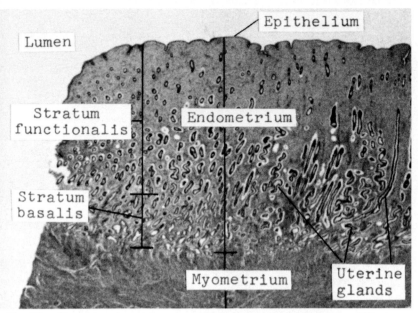

Fig. 88a   Uterus, endometrium x.s. x20. Proliferative phase, 7th - 15th day.

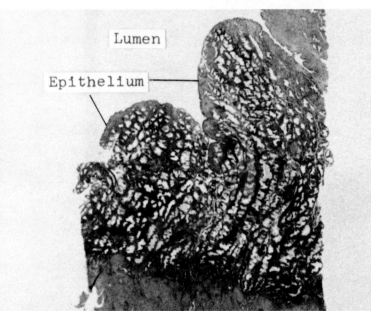

Fig. 88b   Uterus, endometrium x.s. x20. Secretory phase, 16th - 27th day.

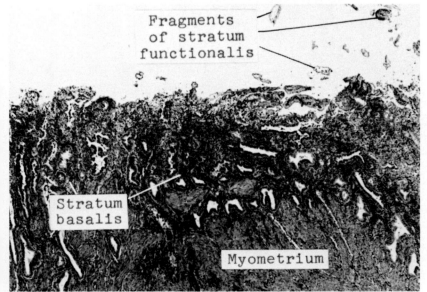

Fig. 88c   Uterus, endometrium x.s. x40. Menstrual phase, 1st - 5th day. The entire stratum functionalis degenerates and is shed. The stratum basalis will give rise to the new stratum functionalis in the next cycle.

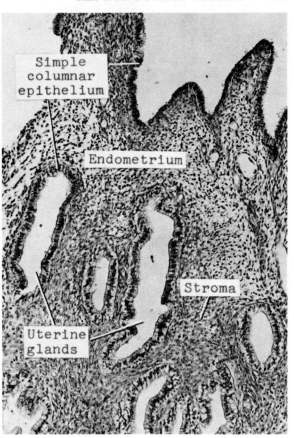

Fig. 88d   Uterus, endometrium x.s. x100. Proliferative phase. Notice that the uterine glands are relatively straight.

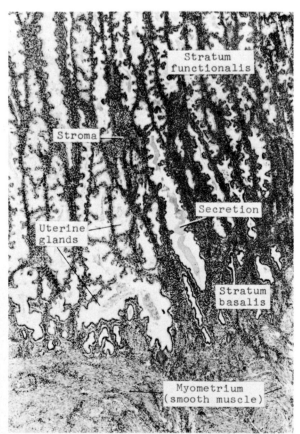

Fig. 88e   Uterus, endometrium x.s. x100. Secretory phase. Notice how the uterine glands are highly coiled and have a sawtooth appearance. The secretion is high in glycogen.

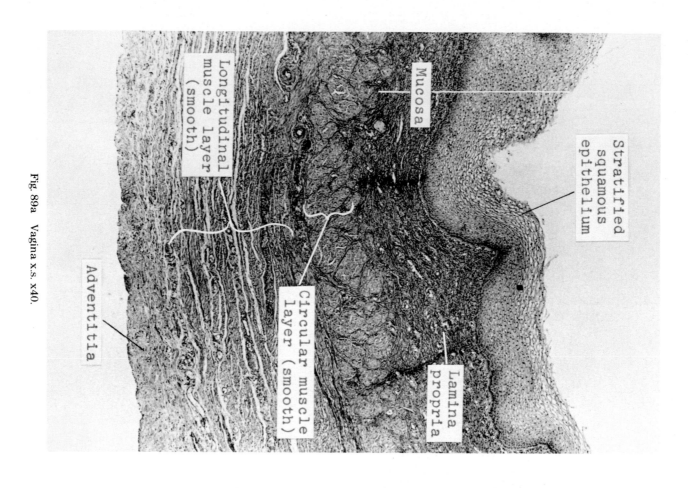

Fig. 89a   Vagina x.s. x40.

Longitudinal muscle layer (smooth)

Mucosa

Stratified squamous epithelium

Adventitia

Circular muscle layer (smooth)

Lamina propria

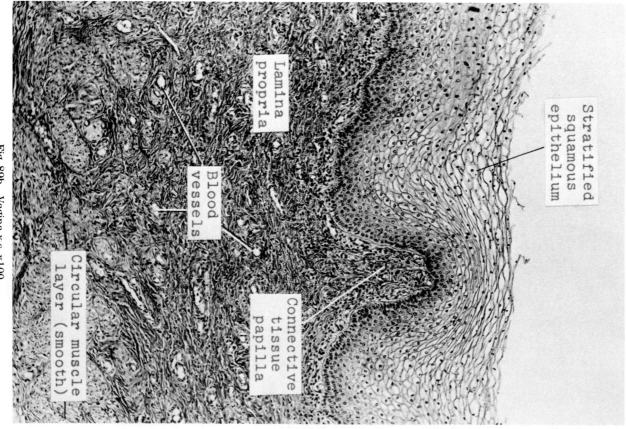

Fig. 89b   Vagina x.s. x100.

Lamina propria

Blood vessels

Stratified squamous epithelium

Circular muscle layer (smooth)

Connective tissue papilla

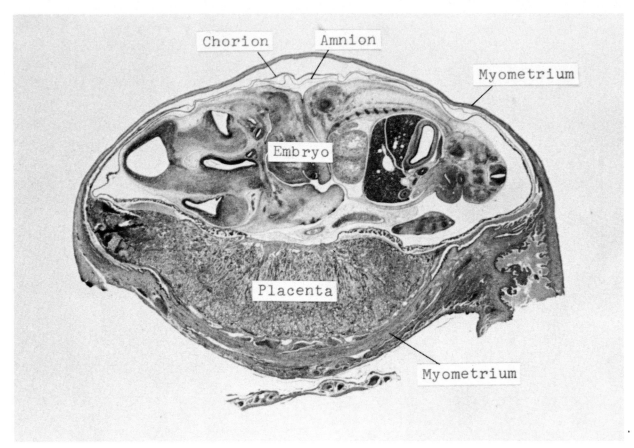

Fig. 90a   Mammalian embryo and placenta x.s. x40.

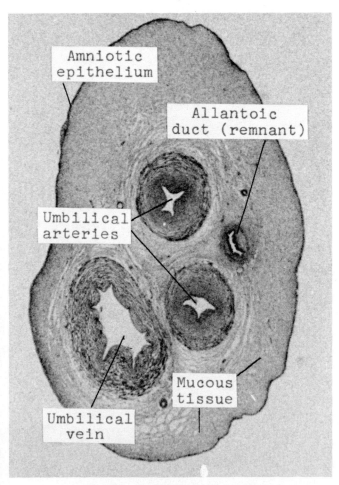

Fig. 90b   Umbilical cord x.s. x10.

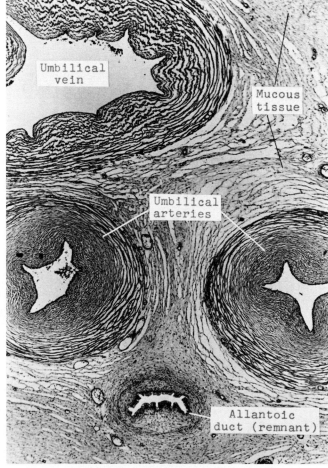

Fig. 90c   Umbilical cord x.s. x40.

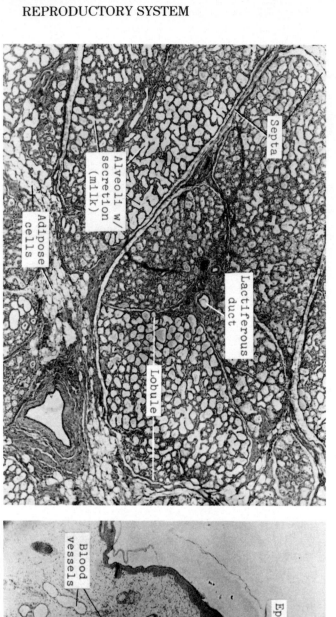

Fig. 91a   Mammary gland (human), non-pregnant x.s. x40.

Lactiferous duct

Lobule

Interlobular collagenous connective tissue

Adipose cells

Glandular alveoli

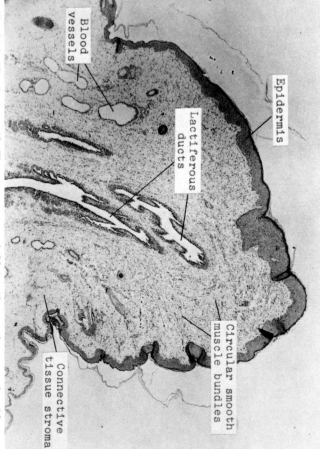

Fig. 91c   Mammary gland (human), lactating x.s. x40.

Septa

Alveoli w/ secretion (milk)

Lactiferous duct

Lobule

Adipose cells

Fig. 91b   Mammary gland (human), pregnant x.s. x40. Note the proliferation of the alveoli in each lobule.

Septa

Glandular alveoli

Lactiferous duct

Lobule

Interlobular collagenous connective tissue

Fig. 91d   Nipple l.s. x40. There are around 30 lactiferous ducts, each with an opening on the nipple's surface.

Blood vessels

Lactiferous ducts

Epidermis

Circular smooth muscle bundles

Connective tissue stroma

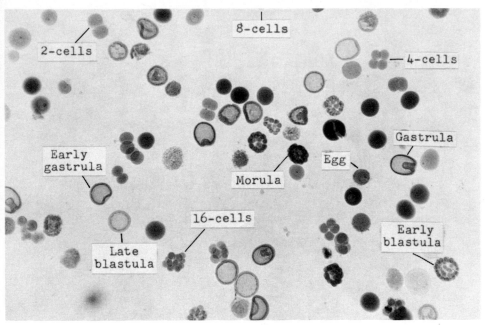

Fig. 92a   Starfish embryos, various stages w.m. x40.

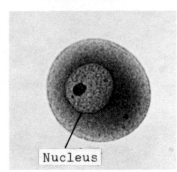

Fig. 92b   Single egg cell.

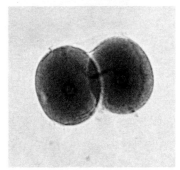

Fig. 92c   Two cell stage.

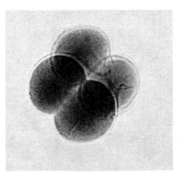

Fig. 92d   Four cell stage.

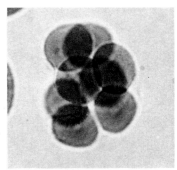

Fig. 92e   Eight cell stage.

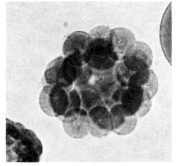

Fig. 92f   Morula.

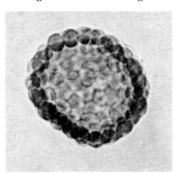

Fig. 92g   Early blastula.

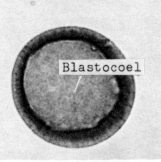

Fig. 92h   Late blastula.

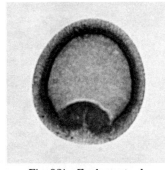

Fig. 92i   Early gastrula.

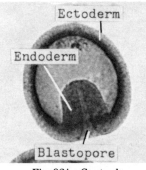

Fig. 92j   Gastrula.

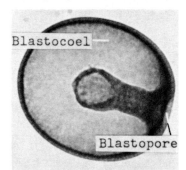

Fig. 92k   Gastrula.

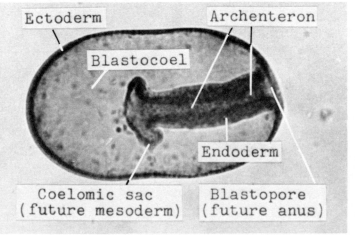

Fig. 92l   Gastrula w.m. x100.

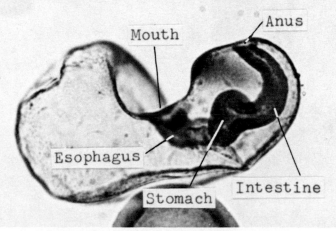

Fig. 92m   Bipinnaria larva of starfish w.m. x100.

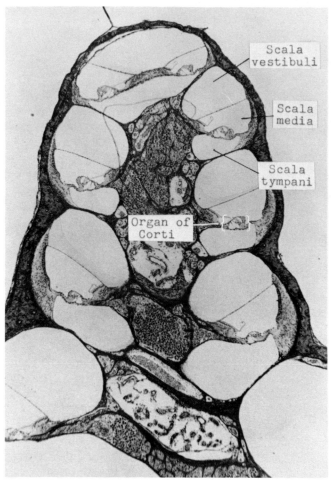

Fig. 93a   Cochlea l.s. x40.

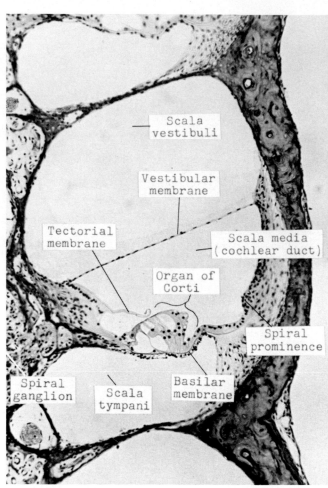

Fig. 93b   Cross section through one of the turns of the cochlea x.100.

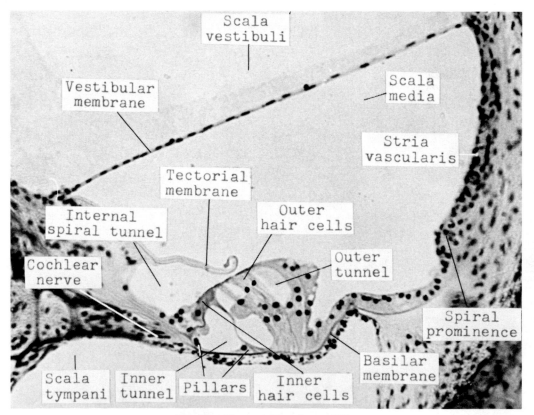

Fig. 93c   Organ of Corti x.s. x200.

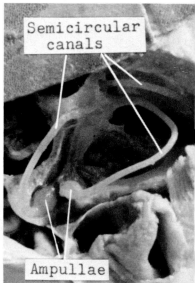

Fig. 93d   Semicircular canals and ampullae of a shark x1.

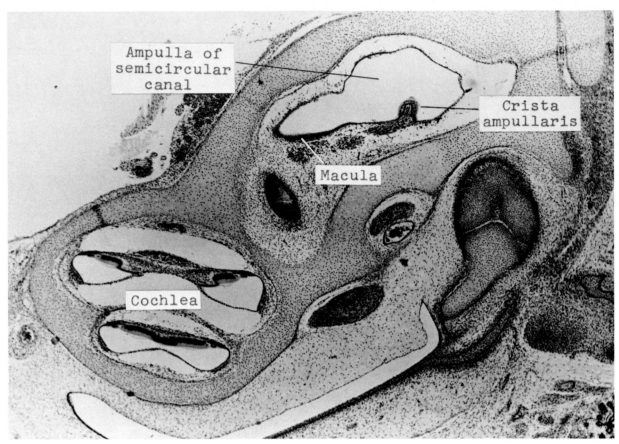

Fig. 94a   Crista ampullaris x.s. x40.

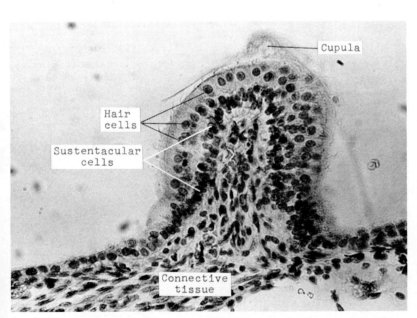

Fig. 94b   Crista ampullaris x.s. x430. The cupula is moved by the movement of the endolymph. The cupula's motion is sensed by the hair cells.

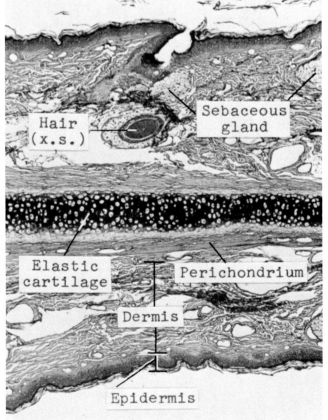

Fig. 94c   Auricle (pinna) l.s. x40. (Also see Figs. 8c & 8d.)

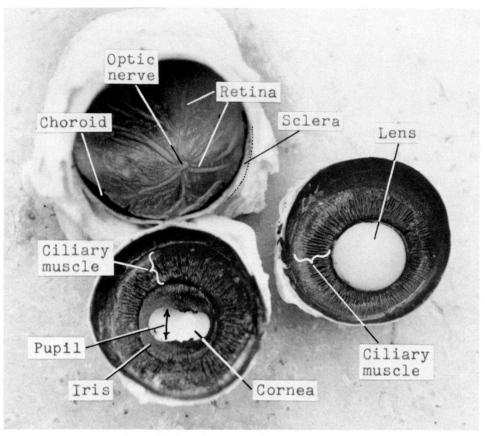

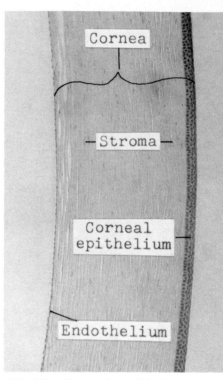

Fig. 95a   Dissected sheep eyes x.s. The two left halves are of the same eye with the lens removed to show the pupil. The right eye shows the lens in place.

Fig. 95b   Cornea l.s. x430. The cornea is made of bundles of transparent collagen fibers arranged in thin lamellae.

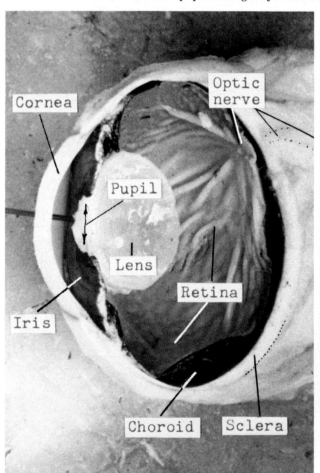

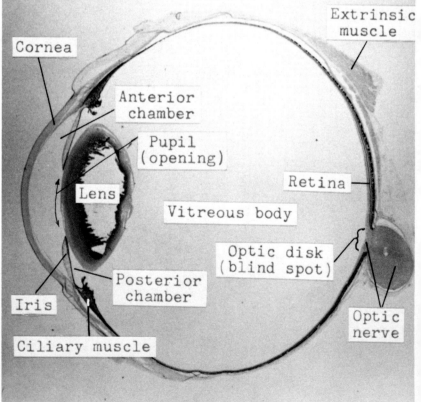

Fig. 95c   Sheep eye dissection l.s.

Fig. 95d   Monkey eye l.s. x5.

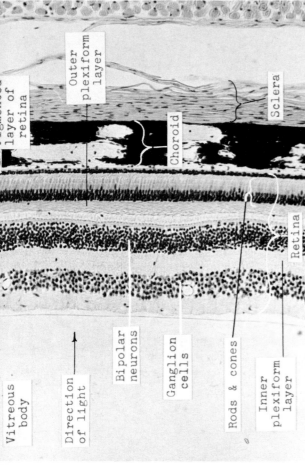

Vitreous body

Direction of light

…mented layer of retina

Outer plexiform layer

Choroid

Sclera

Bipolar neurons

Ganglion cells

Rods & cones

Inner plexiform layer

Retina

Fig. 96b   Retina of monkey eye l.s. x430. Notice that light must pass through several layers of nerve cells before it reaches the light-sensitive rods & cones. (Rods & cones are neurons also.)

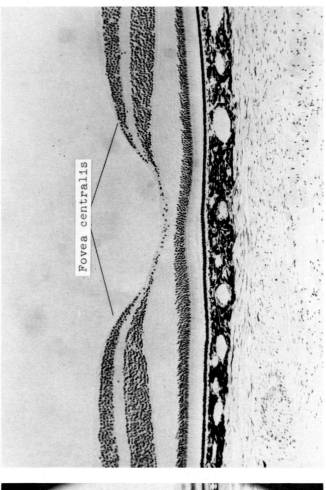

Fovea centralis

Fig. 96d   Fovea centralis of monkey eye l.s. x100. Your sharpest vision comes from this area because: (a) the fovea contains only cones in high concentration and (b) the reduction in overlying layers of other cells.

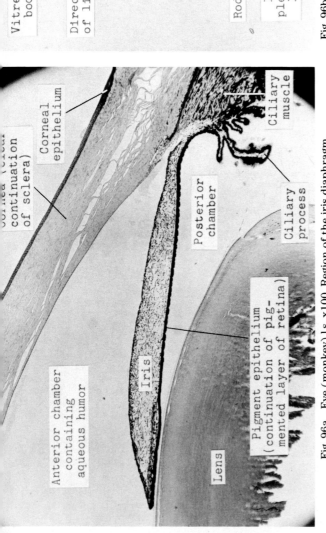

Corneal epithelium

Cornea (clear continuation of sclera)

Ciliary muscle

Posterior chamber

Ciliary process

Anterior chamber containing aqueous humor

Iris

Pigment epithelium (continuation of pig- mented layer of retina)

Lens

Fig. 96a   Eye (monkey) l.s. x100. Region of the iris diaphragm.

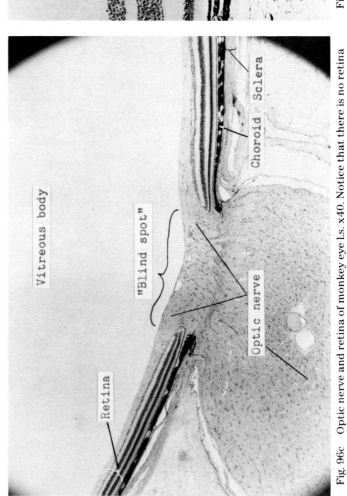

Vitreous body

Choroid   Sclera

"Blind spot"

Retina

Optic nerve

Fig. 96c   Optic nerve and retina of monkey eye l.s. x40. Notice that there is no retina from where the optic nerve enters the posterior part of the eye. Since there are no rods or cones in this area, no light is perceived and it is known as the blind spot. Axons of ganglion cells form the optic nerve.

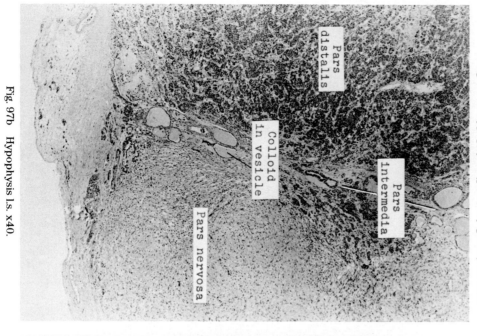

Fig. 97a Hypophysis (pituitary gland) l.s. x10.

Pars distalis

Pars intermedia

Capsule

Pars nervosa

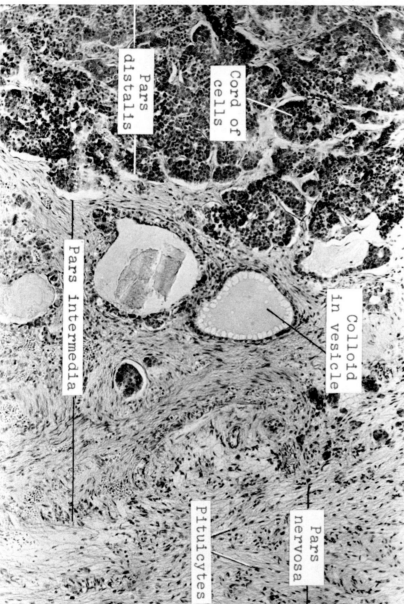

Fig. 97b Hypophysis l.s. x40.

Pars distalis

Colloid in vesicle

Pars intermedia

Pars nervosa

Fig. 97c Hypophysis l.s. x100.

Pars distalis

Cord of cells

Pars intermedia

Colloid in vesicle

Pars nervosa

Pituicytes

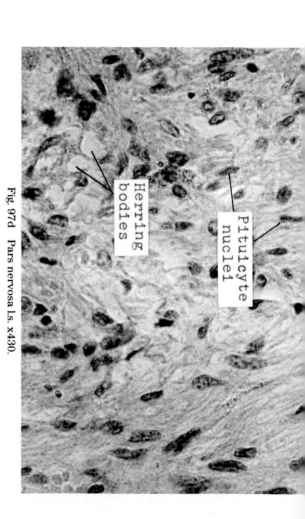

Fig. 97d Pars nervosa l.s. x430.

Herring bodies

Pituicyte nuclei

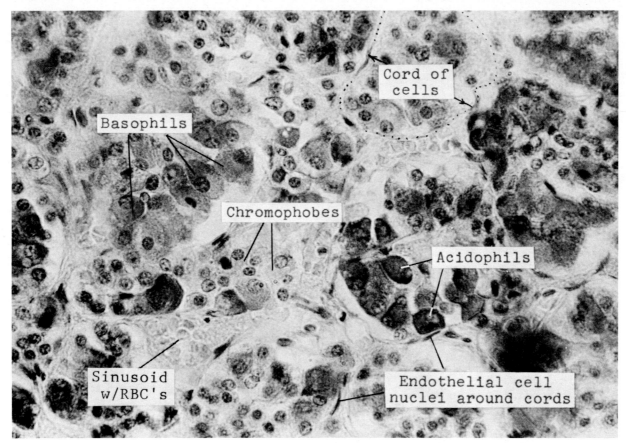

Fig. 98a   Pars distalis l.s. x430.

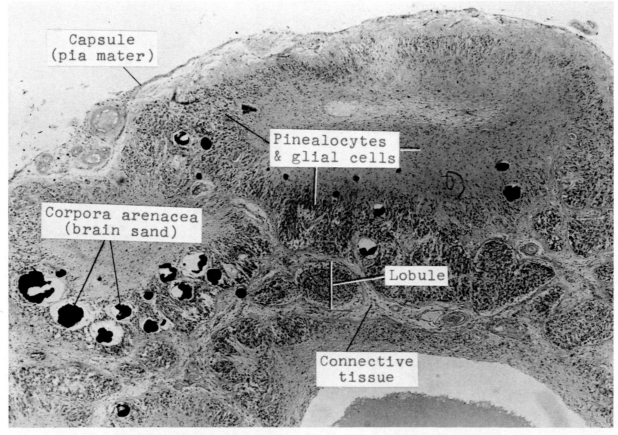

Fig. 98b   Pineal gland l.s. x40. The amount of brain sand increases with age.

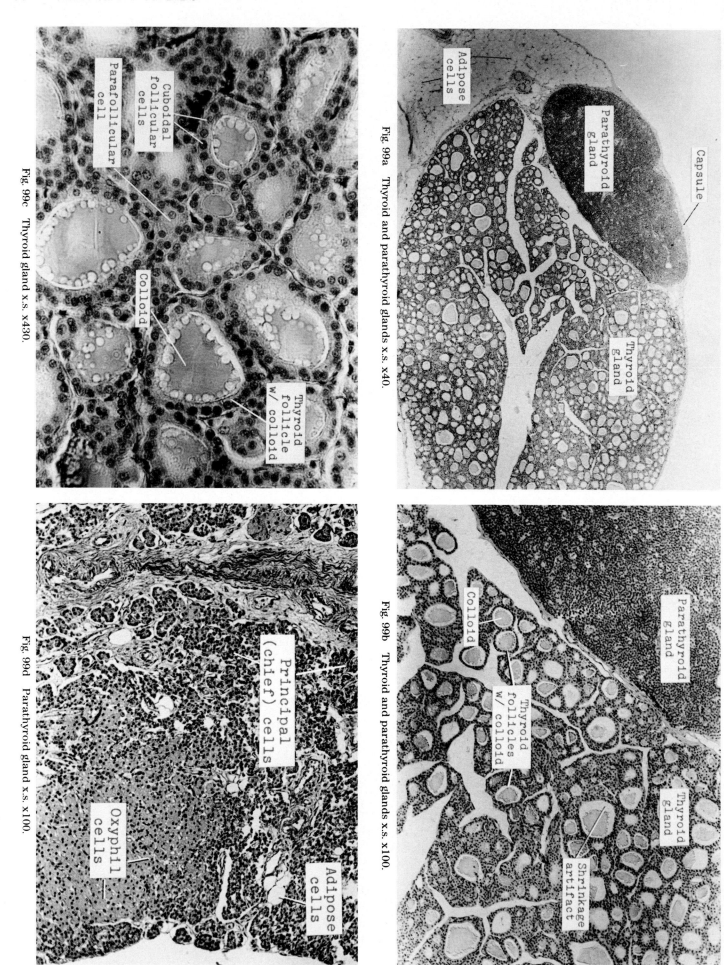

Parafollicular cell

Cuboidal follicular cells

Colloid

Thyroid follicle w/ colloid

Fig. 99c Thyroid gland x.s. x430.

Adipose cells

Parathyroid gland

Capsule

Thyroid gland

Fig. 99a Thyroid and parathyroid glands x.s. x40.

Principal (chief) cells

Oxyphil cells

Adipose cells

Fig. 99d Parathyroid gland x.s. x100.

Colloid

Thyroid follicles w/ colloid

Parathyroid gland

Thyroid gland

Shrinkage artifact

Fig. 99b Thyroid and parathyroid glands x.s. x100.

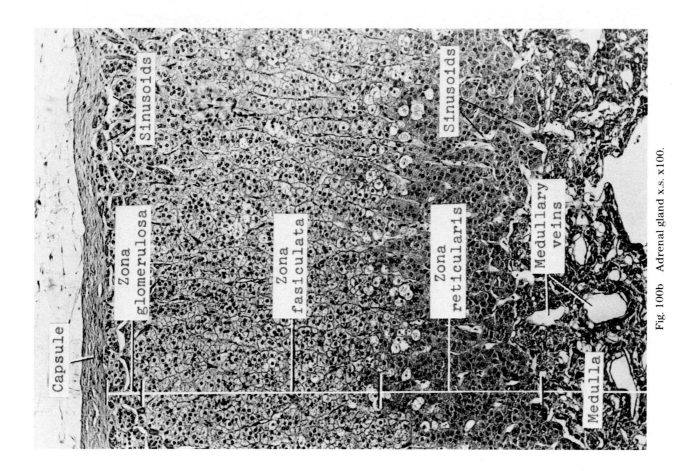

Capsule

Sinusoids

Sinusoids

Zona glomerulosa

Zona fasiculata

Zona reticularis

Medullary veins

Medulla

Fig. 100b   Adrenal gland x.s. x100.

Capsule

Medullary veins

Cortex

Medulla

Cortex

Fig. 100a   Adrenal gland x.s. x40.

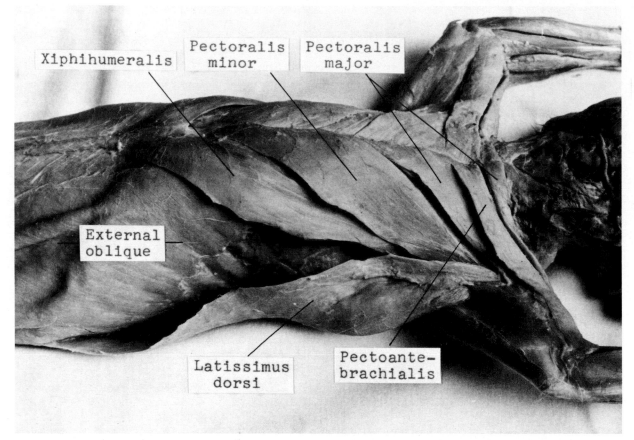

Fig. 101a   Pectoral muscles. (Dissections by Dr. John Lyon.)

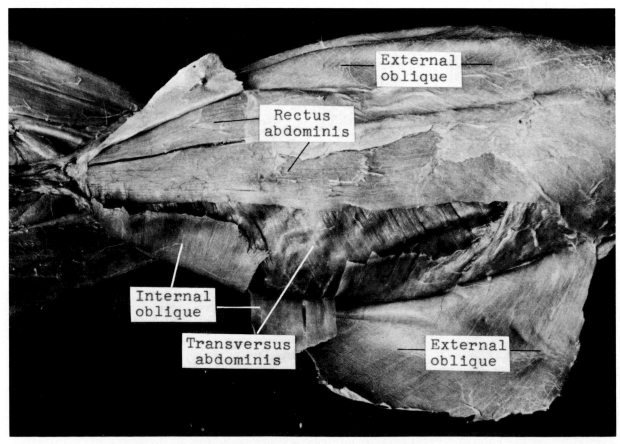

Fig. 101b   Abdominal muscles, ventral view.

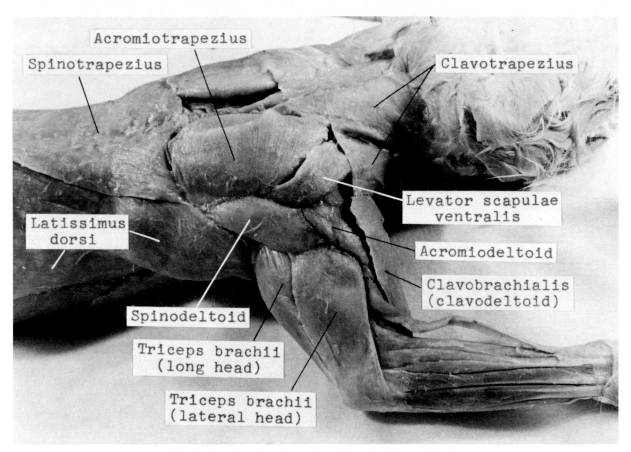

Fig. 102a   Superficial muscles of the shoulder and back.

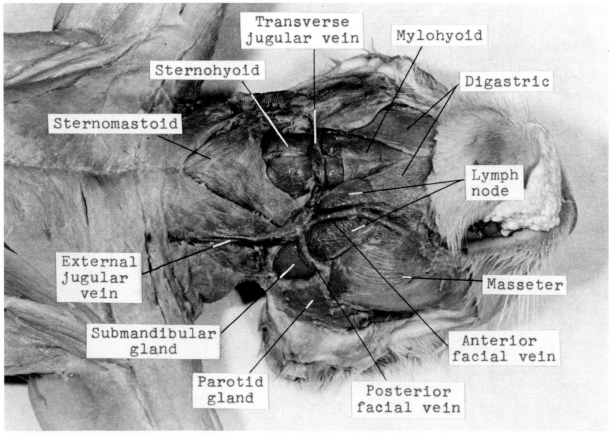

Fig. 102b   Superficial neck and jaw muscles.

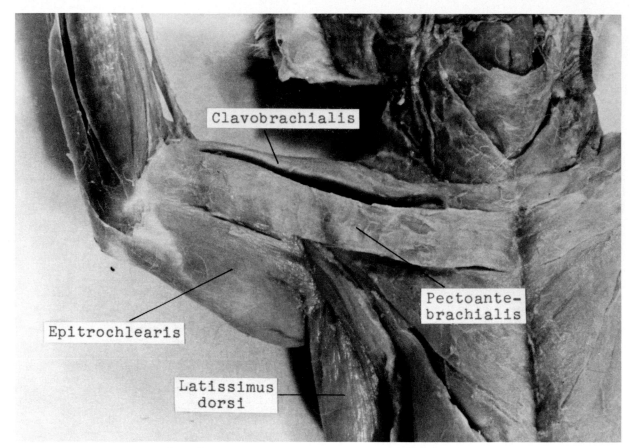

Fig. 103a   Superficial upper arm muscles, medial surface.

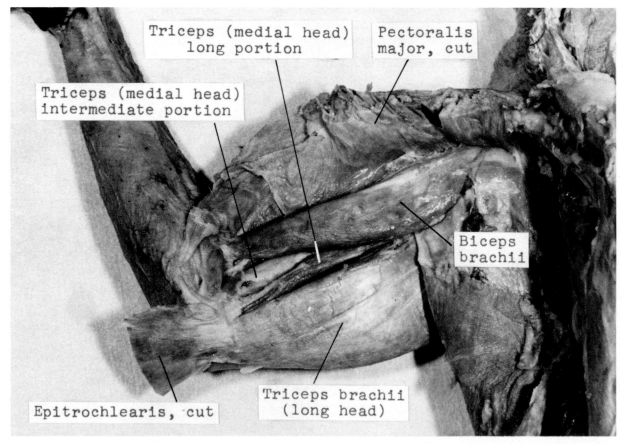

Fig. 103b   Deep upper arm muscles, medial surface.

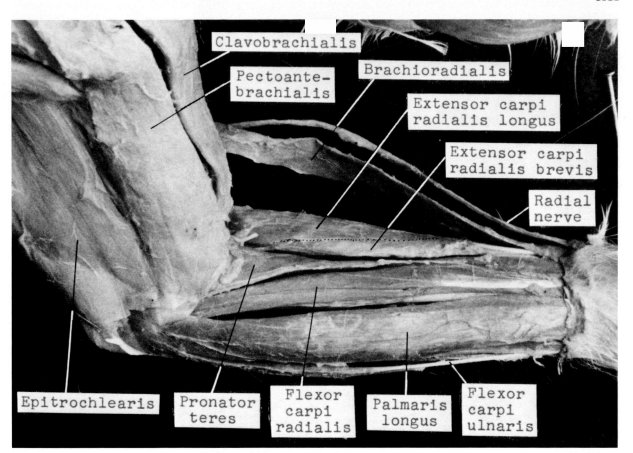

Fig. 104a   Muscles of the forearm, medial surface.

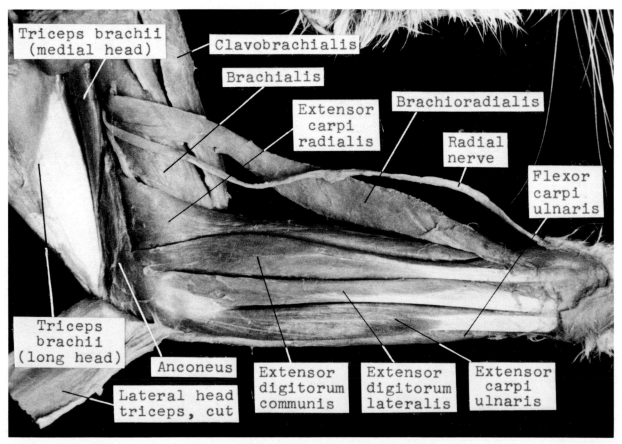

Fig. 104b   Muscles of the forearm, lateral surface.

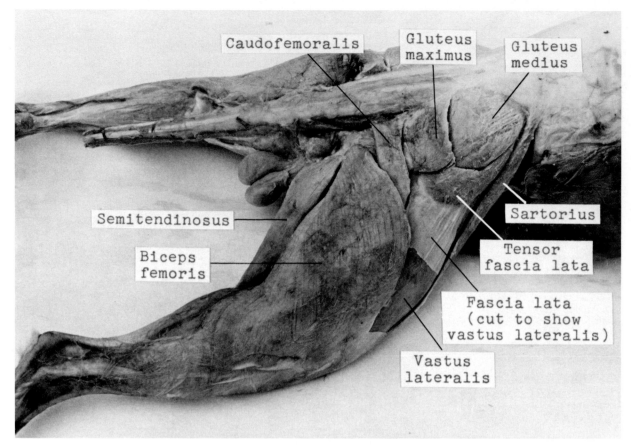

Fig. 105a   Superficial muscles of the thigh, lateral view.

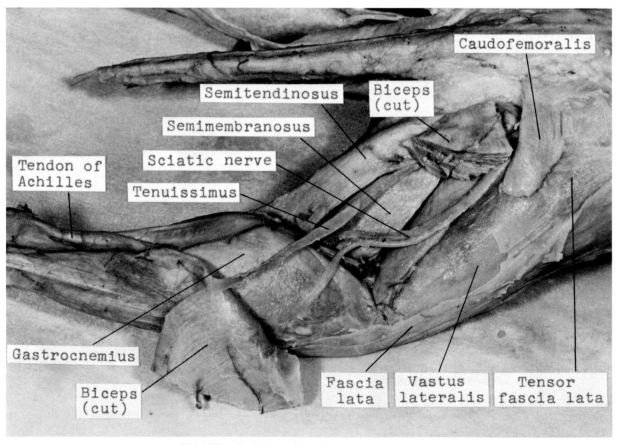

Fig. 105b   Deep muscles of the thigh, lateral view.

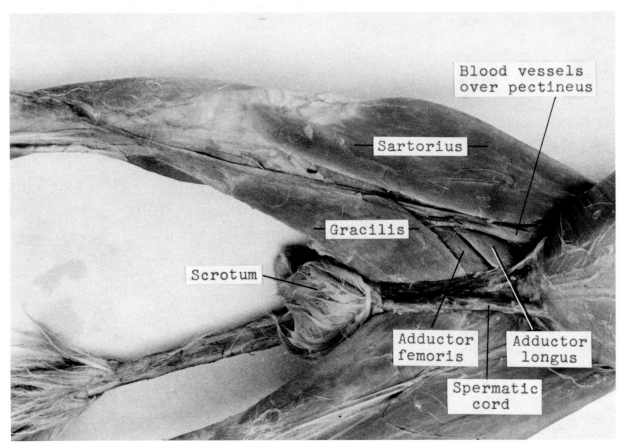

Fig. 106a   Superficial muscles of the thigh, medial view.

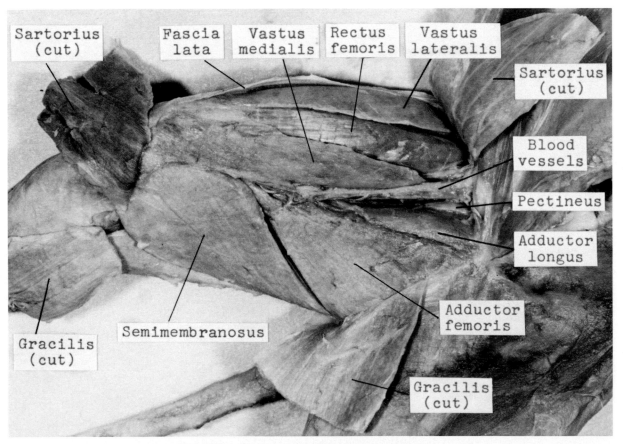

Fig. 106b   Deep muscles of the thigh, medial view.

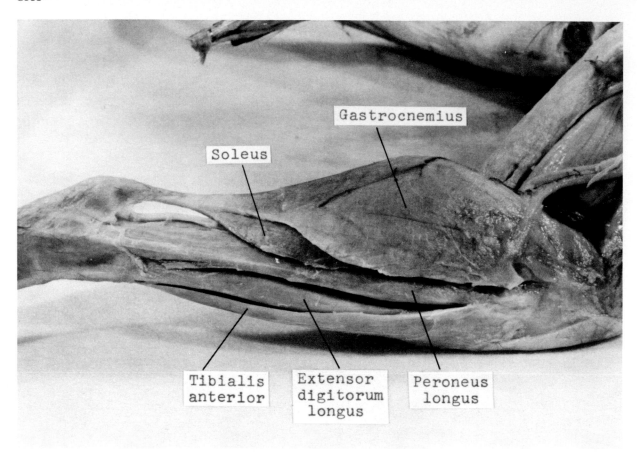

Fig. 107a   Muscles of the lower leg, lateral view.

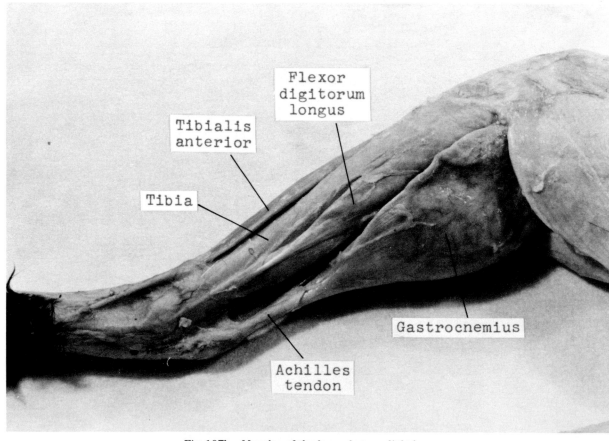

Fig. 107b   Muscles of the lower leg, medial view.

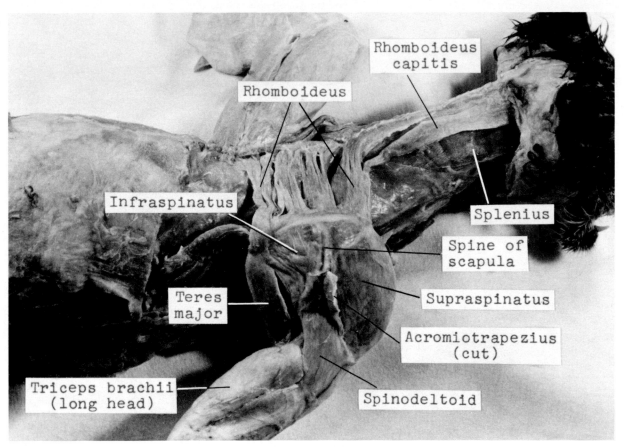

Fig. 108a   Deep muscles of the back and scapula, dorso-lateral view.

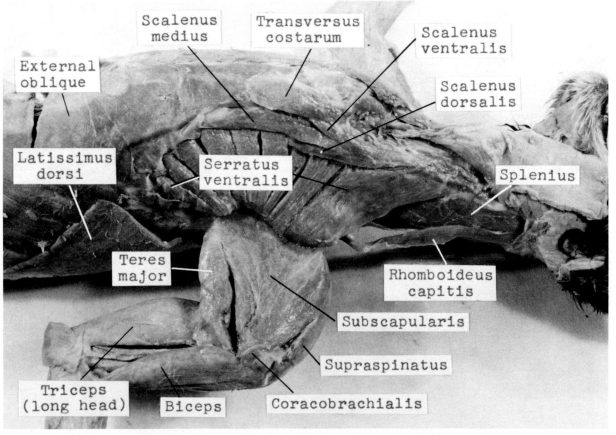

Fig. 108b   Deep muscles of the chest and scapula, medial surface.

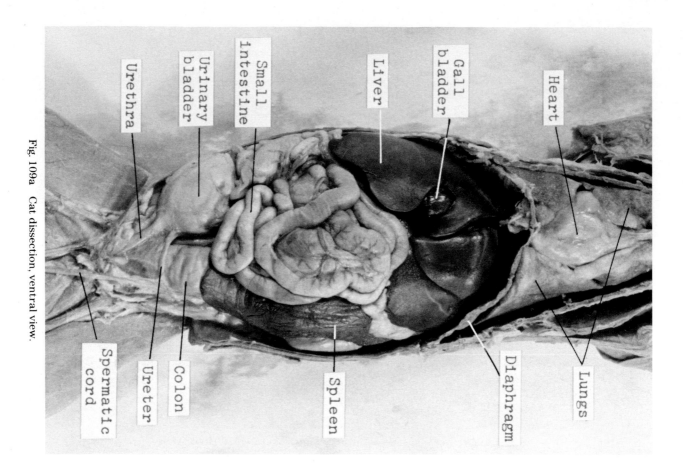

Fig. 109a   Cat dissection, ventral view.

Urethra

Urinary bladder

Small intestine

Liver

Gall bladder

Heart

Spermatic cord

Ureter

Colon

Spleen

Diaphragm

Lungs

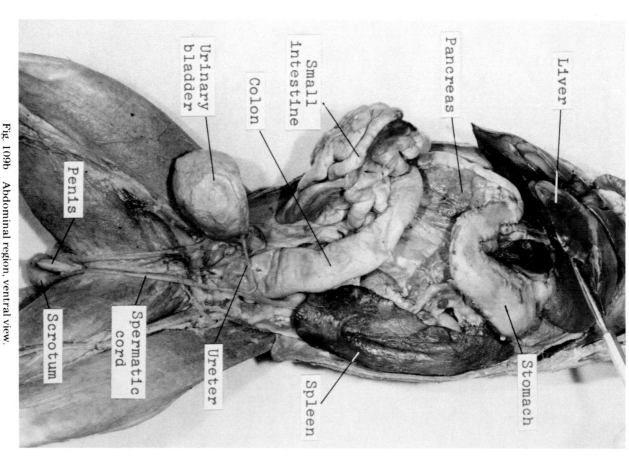

Fig. 109b   Abdominal region, ventral view.

Penis

Urinary bladder

Colon

Small intestine

Pancreas

Liver

Scrotum

Spermatic cord

Ureter

Spleen

Stomach

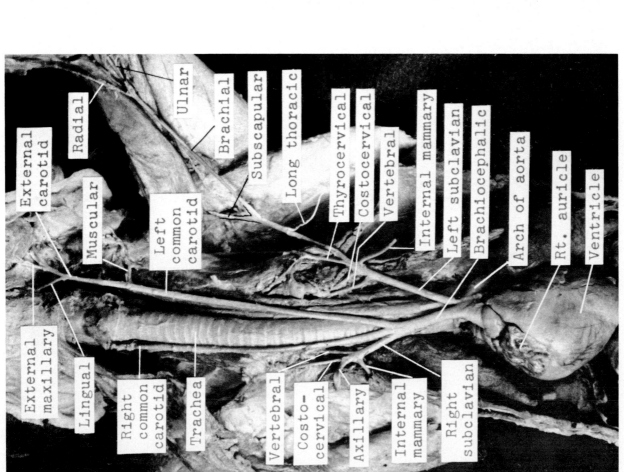

Fig. 110b   Arteries of the abdomen and legs.

Fig. 110b labels:
Spleen, Left gastric, Splenic, Adrenolumbar, Ovarian/testicular not shown, Renal, Abdominal aorta, Iliolumbar, External iliac, Superior gluteal, Inferior gluteal, Femoral, Muscular branch, Deep femoral, Hepatic, Celiac, Superior mesenteric, Ureter, Inferior mesenteric, Internal iliac, Caudal, Colon

Fig. 110a   Arteries of the thorax and neck regions.

Fig. 110a labels:
External carotid, Radial, Ulnar, Brachial, Subscapular, Long thoracic, Thyrocervical, Costocervical, Vertebral, Internal mammary, Left subclavian, Brachiocephalic, Arch of aorta, Rt. auricle, Ventricle, Muscular, Left common carotid, External maxillary, Lingual, Right common carotid, Trachea, Vertebral, Costo-cervical, Axillary, Internal mammary, Right subclavian

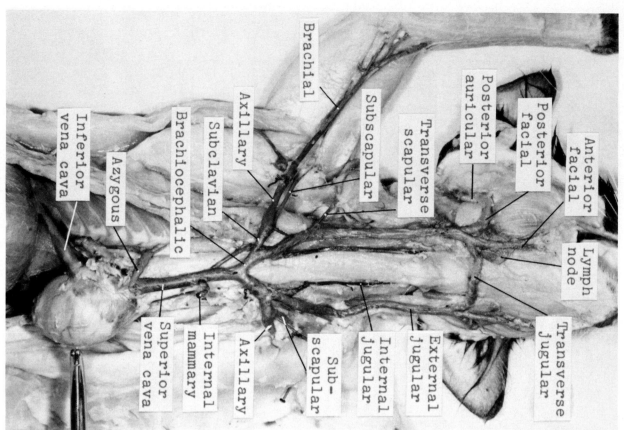

Fig. 111a Veins of the thoracic and neck regions.

Brachial

Subscapular

Transverse
scapular

Posterior
auricular

Posterior
facial

Anterior
facial

Lymph
node

Transverse
jugular

External
jugular

Internal
jugular

Sub-
scapular

Axillary

Axillary

Subclavian

Brachiocephalic

Azygous

Inferior
vena cava

Internal
mammary

Superior
vena cava

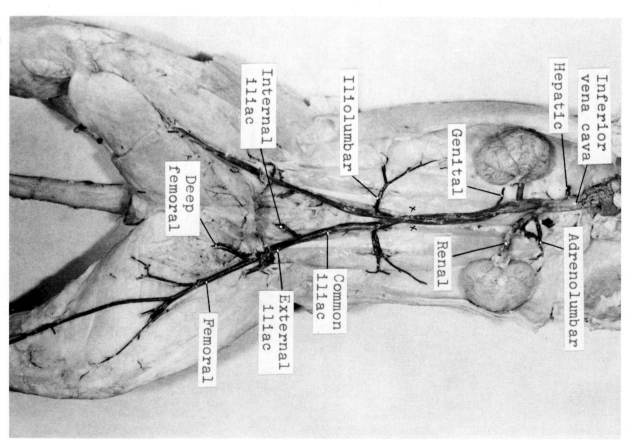

Fig. 111b Veins of the abdomen and legs. (The iliolumbar veins are usually found at the location marked "x.")

Internal
iliac

Iliolumbar

Genital

Hepatic

Inferior
vena cava

Adrenolumbar

Renal

Common
iliac

External
iliac

Deep
femoral

Femoral

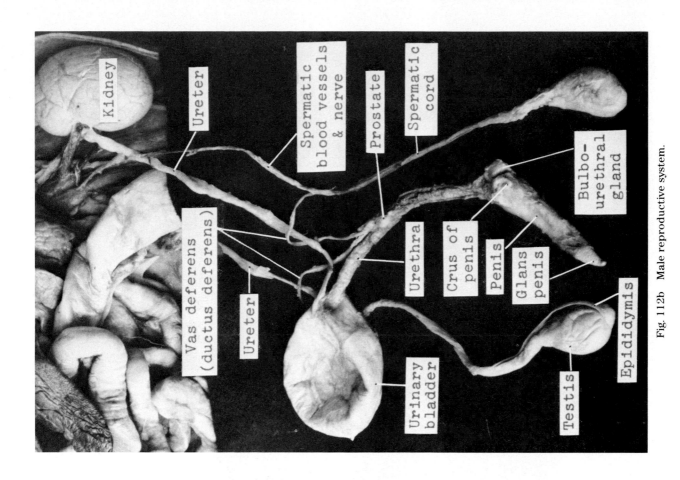

Fig. 112b   Male reproductive system.

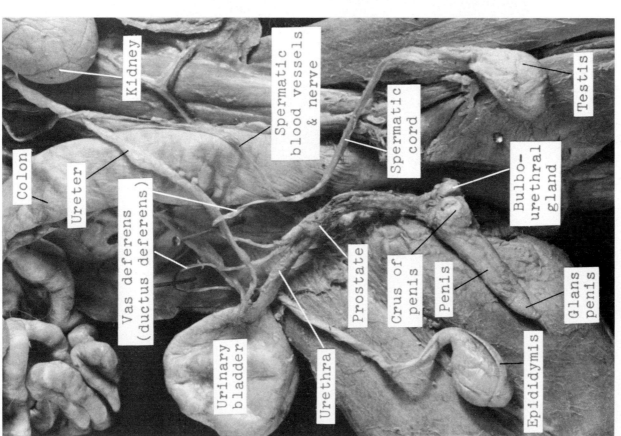

Fig. 112a   Male reproductive system.

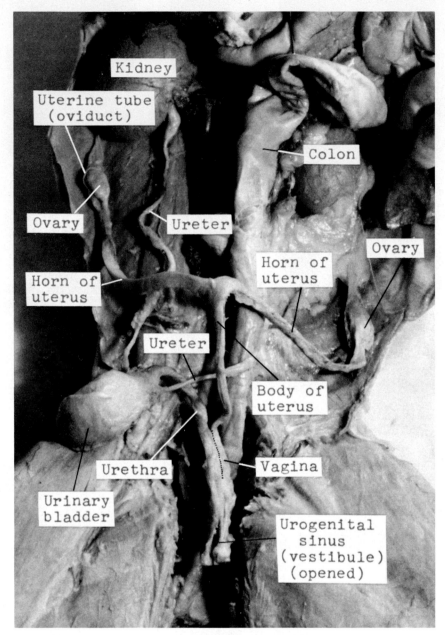

Fig. 113   Female reproductive system.

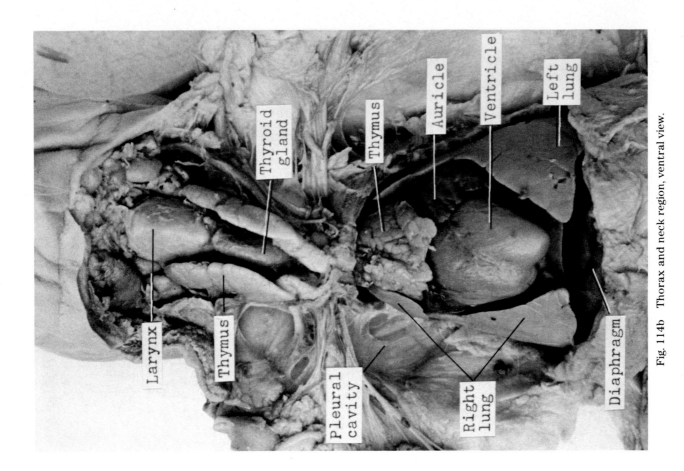

Larynx

Thyroid gland

Thymus

Pleural cavity

Right lung

Auricle

Ventricle

Left lung

Diaphragm

Fig. 114b Thorax and neck region, ventral view.

Larynx

Thymus

Thyroid gland

Auricle

Ventricle

Left lung

Right lung

Liver

Spleen

Small intestine

Umbilical arteries

Urinary bladder

Fig. 114a Fetal pig dissection, ventral view.

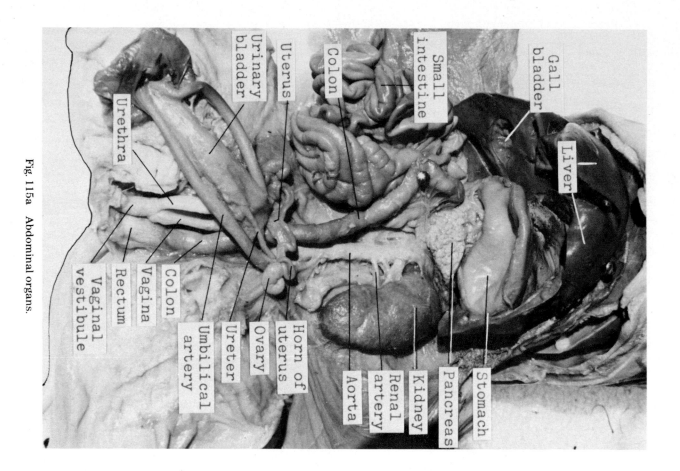

Fig. 115a Abdominal organs.

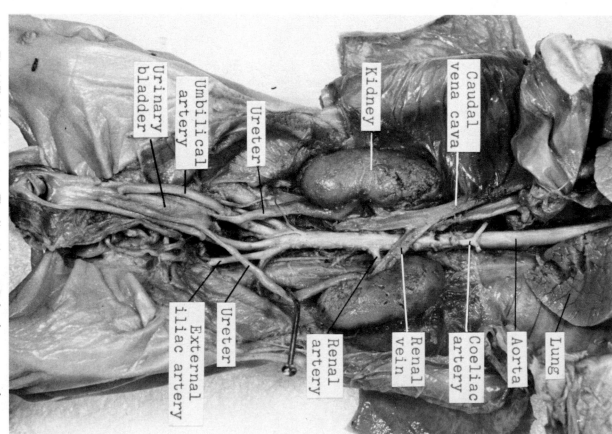

Fig. 115b Urinary system. The digestive system has been removed.

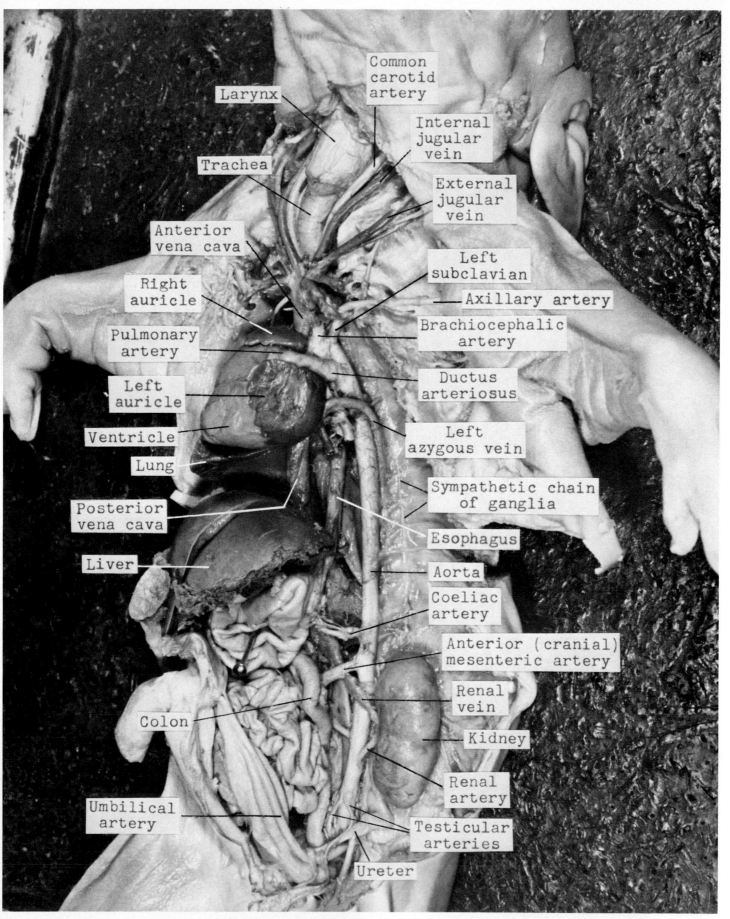

Fig. 116 Fetal pig, ventral view. (Dissection courtesy of Stephen Davenport.)

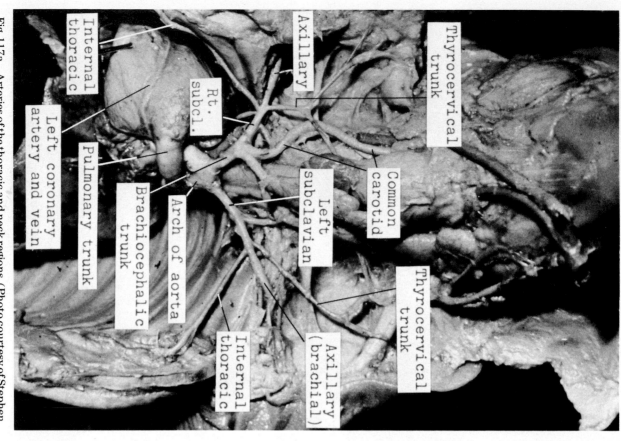

Fig. 117a Arteries of the thoracic and neck regions. (Photo courtesy of Stephen Davenport.)

Internal thoracic

Left coronary artery and vein

Pulmonary trunk

Brachiocephalic trunk

Arch of aorta

Rt. subcl.

Axillary

Left subclavian

Common carotid

Thyrocervical trunk

Internal thoracic

Axillary (brachial)

Thyrocervical trunk

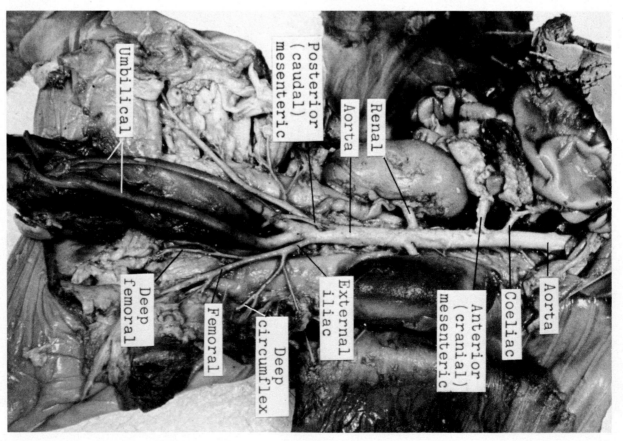

Fig. 117b Arteries of the abdomen and legs.

Umbilical

Posterior (caudal) mesenteric

Aorta

Renal

Deep femoral

Femoral

Deep circumflex

External iliac

Anterior (cranial) mesenteric

Coeliac

Aorta

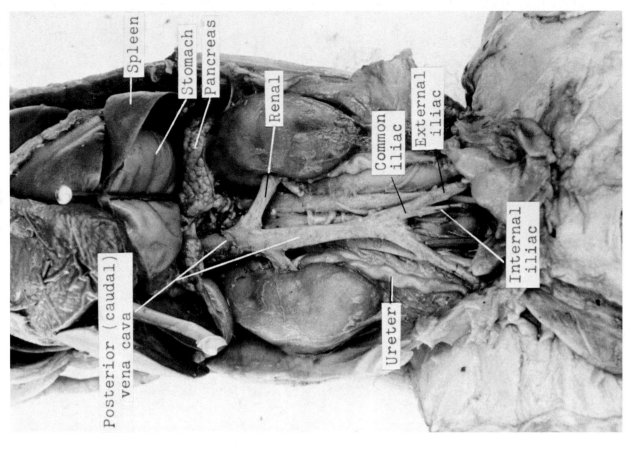

Fig. 118b   Veins of the abdomen.

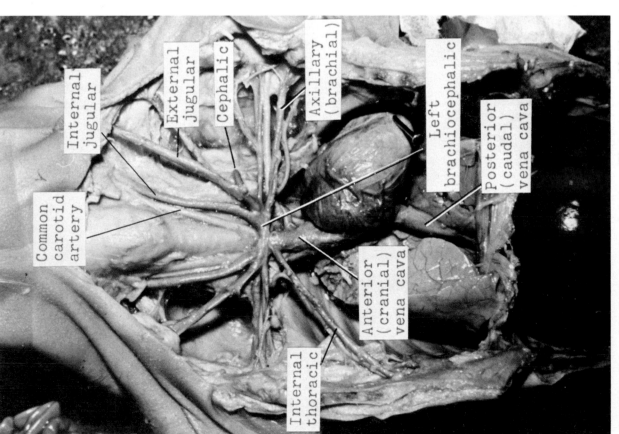

Fig. 118a   Veins of the thoracic and neck regions. (Photo courtesy of Stephen Davenport.)

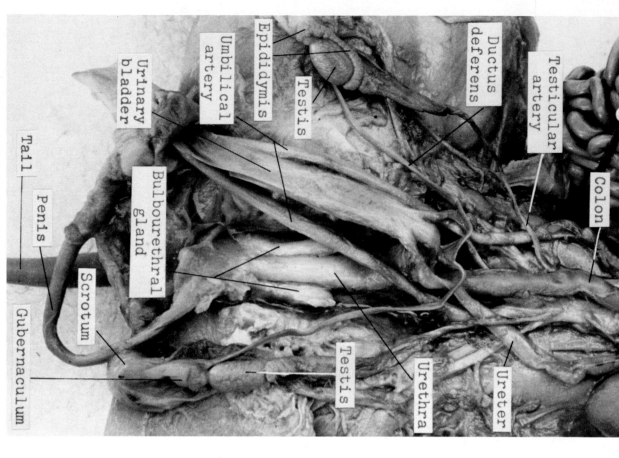

Fig. 119a   Male reproductive system. The spermatic cord (containing the ductus deferens, testicular blood vessels and testicular nerve) has been separated to show the individual components.

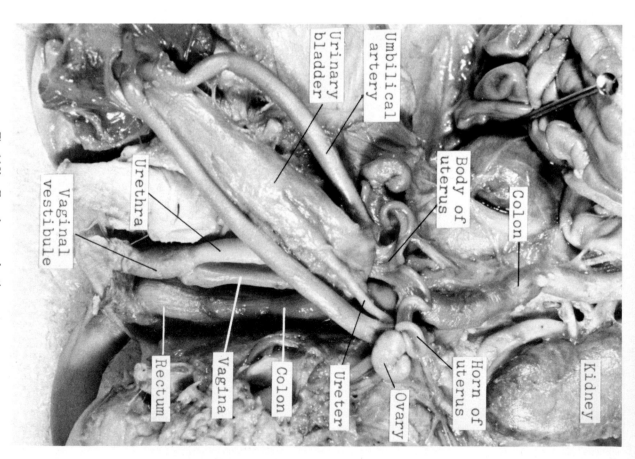

Fig. 119b   Female reproductive system.